Juliardnas Rigamont Dos Reis
Dionne Monteiro
Ana Cássia Ferreira

VIRTUAL USE IN BIOLOGY

Juliardnas Rigamont Dos Reis
Dionne Monteiro
Ana Cássia Ferreira

VIRTUAL USE IN BIOLOGY

SIMULATOR IN VIRTUAL REALITY ENVIRONMENT FOR PLASMATIC MEMBER TEACHING

ScienciaScripts

Imprint
Any brand names and product names mentioned in this book are subject to trademark, brand or patent protection and are trademarks or registered trademarks of their respective holders. The use of brand names, product names, common names, trade names, product descriptions etc. even without a particular marking in this work is in no way to be construed to mean that such names may be regarded as unrestricted in respect of trademark and brand protection legislation and could thus be used by anyone.

Cover image: www.ingimage.com

This book is a translation from the original published under ISBN 978-620-2-55880-8.

Publisher:
Sciencia Scripts
is a trademark of
International Book Market Service Ltd., member of OmniScriptum Publishing Group
17 Meldrum Street, Beau Bassin 71504, Mauritius
Printed at: see last page
ISBN: 978-620-2-64258-3

ACKNOWLEDGEMENTS

If you're reading this page it's because I did it! And it wasn't easy getting this far. From the selection process, through the approval to the conclusion of the Master's degree, it was a long journey and I thank everyone who helped me along this path.

First of all, I would like to thank my husband for the support and encouragement he gave me throughout the Masters, especially in the difficulties faced during this journey.

I thank my daughter Giovana Valentina for her understanding and partnership since her pregnancy, allowing me to fulfill my academic commitments and always in her company.

To my parents who have always excelled at my education. Thank you, painho and mainha, for offering me the opportunity to study, always be present, even living in another state.

I would like to thank the team at the LabITE (Laboratory of Innovation in Teaching Technologies) and LAAI (Laboratory of Applied Artificial Intelligence) for their help. In some way, everyone contributed to this work.

I thank Giordanna de Gregoriis for all her support and patience during Blender's classes.

I thank Francielma Assunção for all her dedication to my project and for the excellent pedagogical tips that made this work great.

To my brother friend Edilson Neri Junior who is always by my side at every moment of my life, supporting, encouraging, calming me. And, during this master's degree, we laughed, cried, despaired, but we were each other's foundation to finish this postgraduate course. Your friendship is essential in my life.

My friend Lorena Cunha who's been with me since graduation. And, at this stage of the Masters, she was always present, helping me when necessary, reading my texts, pointing out where I should improve; accompanying me in the testing of the Masters product, being my daughter's "babysitter" when I needed it. Thank you very much, my friend, I love you unconditionally.

To the friends that the Masters has given me and that I will carry forever in my heart: Aislan de Paula, Ana Carolina, Angela Alexandre, Glenda Alves, Jessica Brigido and Mayara Maciel. It has been two very intense years and I have always been able to count on your support, I hope that, over the years, our friendship will become even more unified.

The friend Mayara Maciel for all the incentive, tips for improving the texts, conversations at dawn reducing my anxiety crises. Destiny put us a barrier of a few kilometers, but this distance is only physical, because it is always present in my life and was fundamental to the conclusion of this dissertation.

My friend Glenda Lent, owner of the warmest embrace and partner of orientation, with whom I shared anguish, afflictions, joys, conquests. We wiped each other's tears a lot, but we also laughed and had a lot of fun in those two years. Thank you for all your support, help and friendly shoulder.

To my advisor Dionne Cavalcante Monteiro for his partnership in building this work, for all the teaching and patience he had with me during these two years. And may this partnership not end here.

My supervisor Ana Cássia Sarmento Ferreira for sharing her years of experience teaching Cellular Biology, directing the main aspects that the product should present to make the teaching-learning process more enjoyable.

To teachers and staff of the Program for Creativity and Innovation in Higher Education Methodologies (PPGCIMES) for their contribution to my academic training, especially to teacher Dr. Cristina Vaz for her welcome and teaching.

To undergraduate students of the UEPA biomedicine course, and of Biological Sciences of UFPA and IFPA. And to the graduate students of PROFBIO/UFPA for their willingness to participate in the product testing process of this dissertation. I hope my product has contributed in some way to your teaching-learning process.

"I listen, I forget.
I see, I remember.
I interact, I understand."

Chinese proverb

SUMMARY

This dissertation presents the process of creation, development, testing and results of a simulator in a Virtual Reality (VR) environment for the plasma membrane and was developed as a didactic resource to assist the teaching-learning relationships of the subject Cellular and Molecular Biology in higher education. The methodology used was the experimental scientific method, following the phases of observation, problem elaboration, hypothesis survey, experimentation, results analysis and conclusion. The *software* was developed by an interdisciplinary team and the agile methodology was used. As steps in the process of building the product, a bibliographic research was performed, as theoretical basis, followed by the modeling of the plasma membrane and animal cell, in the 3D modeling *software Blender*. Later, the modeling was integrated to the VR environment, through the *Unity* 3D game development engine, culminating in the programming of phospholipid movements, transport of small substances through the plasma membrane and forms of user interaction with the virtual environment. For immersion in the cellular environment in VR, the VR *headset,* HTC VIVE *Óculus,* consisting of motion sensors, interaction controls and immersion glasses, is used. The product was tested with professors of Cellular and Molecular Biology from the University of the State of Pará (UEPA), the Federal Institute of Education, Science and Technology of Pará (IFPA) and the Federal University of Pará (UFPA). With undergraduates from the UFPA Biomedicine course, from the IFPA and UFPA Biological Sciences/Licentiature course, and with postgraduates from the Graduate Program in Biology Teaching (PROFBIO) at UFPA. The test data were obtained through questionnaires after contact and use of the VR simulator. From the content analysis, it was perceived that the simulator was considered an important tool that contributed/contributed to facilitate the learning of the content related to the structural and functional aspects of the plasma membrane, besides making this learning playful and enjoyable.

Keywords: Molecular Biology. Plasma Membrane. Teaching. Teaching resource. Virtual Reality.

ABSTRACT

This dissertation presents the process of creation, development, testing and results of a Virtual Reality (VR) environment simulator of a plasma membrane and was developed as a didactic resource to assist the teaching-learning relationships of the Cellular and Molecular Biology discipline in higher education. The methodology used was the experimental scientific method, following the observation, problem elaboration, hypothesis raising, experimentation, results analysis and conclusion phases. The software was developed by an interdisciplinary team and made use of agile methodology. As steps of the product construction process, a bibliographic research was carried out, as a theoretical basis, followed by the modeling of the plasma membrane and the animal cell, in the Blender 3D modeling software. Subsequently, the modeling was integrated into the VR environment through the Unity 3D game development engine, culminating in the programming of phospholipid movements, the transport of small substances across the plasma membrane and in forms of user interaction with the game. Virtual environment. For immersion in the cellular environment in VR, the HTC VIVE Óculus VR headset is made up of motion sensors, interaction controls and immersion glasses. The test of the product was conducted with professors of the Cellular and Molecular Biology discipline of the State University of Pará (UEPA), the Federal Institute of Education, Science and Technology of Pará (IFPA) and the Federal University of Pará (UFPA). With undergraduate students of the Biomedicine course at UEPA, the Biological Sciences / Undergraduate course at IFPA and UFPA and postgraduates from the UFPA Postgraduate Program in Biology Teaching (PROFBIO). Test data were obtained through questionnaires after contact and use of the VR simulator. From the content analysis it was realized that the simulator was considered an important tool and contributed to facilitate the learning of content related to the structural and functional aspects of the plasma membrane, besides making this learning fun and enjoyable.

Keywords: Molecular biology. Plasma membrane. Teaching. Didactic resource. Virtual reality.

1. INTRODUCTION

Those who indulge in practice without science are like the navigator who boards a ship without rudder or compass. Practice must always be based on good theory. Before making a case a general rule, try it two or three times and check whether the experiments produce the same effects. No human research can be considered true science if it does not involve demonstrations.

(Leonardo da Vinci)

With the advancement of technology and its constant presence in the daily lives of the population, the classrooms are occupied by a generation increasingly integrated with it, and there is an expectation that the new technologies will bring rapid solutions to the changes in teaching, focused on student learning and not on the transmission of knowledge.

However, despite all the technological innovations, it can be observed that the transmission of knowledge through the expositive method is still recurrent in education systems, in which, most of the times, the students are referred to a demotivating passivity, because, for them, there is nothing left but to "learn" the knowledge that is being passed on. Therefore, it is important to adopt new resources.

The current teaching system remains the same, with desks lined up, blackboards in front, teachers standing talking and students sitting listening, providing totally obsolete learning based on the transmission of information (MORAN, 2012).

Universities that should be an effectively significant and innovative institution are often not very stimulating for teachers and students. They present aged methods and procedures, far from the current technological demands. Most of the time, students attend classes because they are required, not by real choice, interest, motivation, or achievement (MORAN, 2013).

The educator of the 21st century needs to use new approaches in the classroom that allow a better interaction with this generation called "digital natives"[1]. For education can no longer be restricted to the teacher writing on a blackboard and the student just reproducing the knowledge in notebooks and doing book exercises to consolidate his learning, which will culminate in a written test to verify the appropriation of knowledge (FIALHO, 2018).

Higher education teachers have difficulties in attracting students and keeping them focused in the classroom, as they are scattered in the large volume of information available, by

---

[1] Native digital is the term applied to the whole generation that emerged from the beginning of the 21st century, when digital technology began to be inserted on a large scale in various devices and devices that replaced the then analogue technology applied to telephones and televisions.

the different technological means. Therefore, changes are needed in the current teaching-learning process in order to bring it closer to the students' profile, and in the front line of these changes is the teacher (RODRIGUES et al., 2011).

With the proper use of current technologies, universities can become a set of rich spaces for meaningful, face-to-face and digital learning that motivate students to learn actively, to research all the time, to be proactive, to take initiatives and to interact (MORAN, 2013).

Considering the teaching of Cell Biology, it is important to highlight that the study of cells requires a certain degree of abstraction and imagination on the part of teachers and students, because cell structures are microscopic. The advent of the microscope has allowed this microscopic world, previously invisible and unknown, to become visible and perceptible, but the possibility of three-dimensionality of this world requires a greater understanding of what is observed. Thus, pedagogical alternatives are necessary to facilitate its teaching-learning process, however, the most frequently used are the photos, schematics, drawings present in textbooks (PALMERO, 2000).

Teachers cannot limit themselves to mastering only the content, they also need to seek differentiated didactic resources that arouse students' interest and facilitate the teaching-learning process. Among the many didactic resources, VR technology can be a facilitating tool in the teaching-learning process.

RV is an instrument that has been developing more and more in recent years, and has shown a positive effect when associated with education. This resource has features that enable the user to have contact with realistic situations, interacting directly with content, which could hardly be explored in a real scenario (TORI, 2010).

The content of Cell Biology is historically considered difficult to assimilate (PALMERO, 2000). Images even exist, but an illustration, movement in a video or even an animated image on the computer are not enough for students to realize the veracity of what they mean and/or represent. Even these images having a three-dimensional aspect, they are not always able to pass on information adequately. And since it is not possible to handle and visualize such structures in their everyday life, many students conclude their graduation with the concept that these are flat elements and end up reproducing this to their future students (FERNANDES; ZAMA, 2018).

In view of this, it was thought to build a didactic resource in which the student not only visualizes the structures that make up the plasma membrane, but also interact with them and simulate the transport of substances that occur in it.

1.1 Objectives

The general objective is the creation of a simulator in a Virtual Reality environment for teaching the structural and functional aspects of the Plasma Membrane. To achieve this objective, the following specific objectives have been outlined:

✓ Perform bibliographic research from the survey of related works;

✓ Model the structures that make up the plasma membrane and the molecules that cross this structure;

✓ Create the virtual environment that simulates the fluid mosaic model of the plasma membrane;

✓ Simulate the movements of the dynamic structures that make up the plasma membrane;

✓ Simulate passive and active transport through the plasma membrane;

✓ Test the product with teachers of Cellular Biology and with undergraduate students of the Biological Sciences course.

1.2 Justification

The product presented in this dissertation was idealized from discussions of the author of this work with her advisor, having as a guideline the possibility of developing an educational instrument in VR significant for the teaching of Cell Biology, her area of initial training.

After researching and understanding what the VR was about, we opted for the approach of a content taught in the discipline of Cellular Biology, called Plasma Membrane, through this technology. This is because this content is considered abstract and difficult for the student to imagine how the structures that constitute this cellular component are organized (PALMERO, 2000).

The drawings in textbooks and curriculum materials have not facilitated the understanding of the contents taught in Cellular Biology. Faced with this impasse, the search for new didactic approaches is necessary to stimulate the teaching-learning process. It was then seen in VR the possibility of proposing a product that would help to diversify higher education methodologies for this subject, an VR simulator for teaching in the plasma membrane, which allows the student to be immersed within the cell, and can visualize and interact with its structures, becoming an active character in the teaching-learning process.

1.3 Materials and Methods

This section presents the material used and the methods used in the development of the research. The material is listed below and the methods are described in the following items.

a) Materials

As material they were used:

✓ Computer: *Intel Core i5* processor, 3.20 GHz CPU, GTX 750 Ti video card, Operating System: *Windows* 10 64-bit;

✓ *Alienware: Intel Core i7* processor, 2.80 GHz CPU, GTX 1060 graphics card, *Windows 10 64-bit* Operating System

✓ HTC VIVE glasses: *headset*, motion sensors and controls (VIVE, 2018);

✓ Free 3D modeling *software Blender* (Blender Inc., 2018);

✓ *Software Unity 3D* (UNITY, 2019);

✓ Books of Cell Biology (ALBERTS et al., 2017), (COOPER; HAUSMAN, 2007).

Besides these programs and specific equipment for the construction and immersion in the cell environment in VR, its development counted on an interdisciplinary team of professionals, composed by a biologist (author of this work), who modeled the cell structures and directed the construction of the simulator, the processes developed and the types of possible interactions; a *design* professional, who assisted in the modeling performed; and two *Unity 3D* programmers, responsible for exporting the three-dimensional structures from *Blender* to *Unity* and developing the simulator in a VR environment.

b) Methods

The methodology used was the experimental type scientific method. In order to perform it, it was necessary to follow the following steps: observation, problem elaboration, hypothesis survey, experimentation, results analysis and conclusion (VIANNA, 2001).

In the observation, based on personal experiences and on a bibliographic survey, it was observed the difficulty of teachers and students in defining and visualizing more complex terms of Cellular Biology, which cover the content of plasmatic membrane. Therefore, the problem of this research is: how to facilitate the understanding of the "Cell Biology" content, specifically the "plasma membrane", by higher education students? The hypothesis considered in this work is that the use of VR, in the teaching-learning process of such content, can be an innovative way to approach the subject.

Because it is a work anchored in the scientific-experimental methodological process, the stages of experimentation were essential for the construction of the final product.

For the construction of the simulator in VR, it was made use of agile methodologies that were chosen because they are adaptive instead of predictive, that is, they adapt to new factors arising instead of trying to previously analyze everything that can happen during the development (SOARES, 2004).

Agile Software Development methods, as the name implies, involve a set of methodologies that serve to accelerate the pace of *software* development processes (MANIFESTO, 2001).

The main techniques of agile methodology exist: *Extreme Programming* (XP); *Scrum*; *Feature Driven Development* (FDD); *Rapid Application Development* (RAD). Among these four techniques, the one that came closest to the reality of the developed product was *Scrum*, so it was chosen to use it.

This technique is based on iterative and incremental planning. It is concerned with the list of tasks that need to be carried out to develop the product that has been designed. The tasks are distributed among the team and constant tests are performed to evaluate if the development is in accordance with the expected or if changes need to be established (MONTEIRO, 2013).

The tests correspond to the process of execution of a product in order to guarantee that the functionalities are in correct functioning, based on the specifications of the product. The objective is to identify flaws or defects in the product before delivery to the end user (NETO, 2007).

Based on *Scrum* concepts, the development cycle of the VR simulator for the study of plasma membrane followed the model represented in Figure 1.

Figure 1 - Agile method for developing the simulator in VR environment

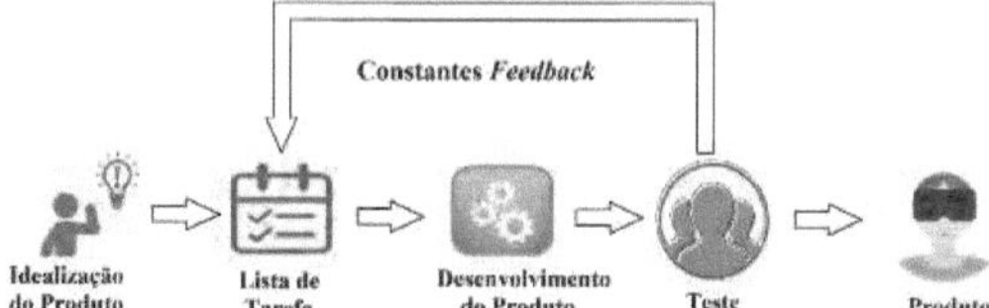

Source: author, 2019.

After the product was designed, the tasks that would be performed during the development of the product were defined, as well as the schedule for the execution of activities. The tasks performed were:

1) Bibliographic survey about virtual reality and the content that the product would be constituted, aiming at a deepening of the theme worked (virtual reality and plasmatic membrane);

2) 3D modeling of the structures that make up the plasma membrane, using the free 3D modeling *software Blender* (Blender Inc., 2018);

3) Integration of modeling to the virtual reality environment through *Unity* 3D (UNITY, 2019), which is an *Integrated Development Environment* (IDE) or Integrated Development Environment, which is *software* that helps in the development process of other *software*, integrating tools and facilitating their use;

4) Programming of the simulator through the C# language, in which ways of user interaction with both structures and the environment were established, enabling the development of the simulator;

5) Test of the tool, which was carried out with undergraduate and graduate students and their respective teachers;

6) Product improvement, which was performed constantly, after each test. Constant *feedback* is an advantage of agile methodologies, and this allows to improve the quality in *software* development;

7) Presentation of the final product.

After the construction of the product, the experimentation stage was carried out with ninety-eight students, ninety-two of them undergraduate and six postgraduate. Undergraduate students attended the Biological Sciences and Biomedicine course, of which thirty-five attended the Biological Sciences/Licentiate course at the Federal Institute of Education, Science and Technology of Pará (IFPA), thirty-five attended the Biological Sciences/Licentiate course at the Federal University of Pará (UFPA), and twenty-two attended the Biomedicine course at the State University of Pará (UEPA). The six master's degree students belong to the Post-Graduate Program in Biology Teaching (PROFBIO) of UFPA. This step consisted in using the product and filling out an evaluation questionnaire. Through this instrument, it was possible to perform the last stages of the scientific method: the analysis of results, followed by the conclusion.

1.4 Work Organization

The text of this dissertation was structured as follows:

✓   **Chapter I:** is intended for the theoretical background that guided the development of the simulator. It deals with VR and Cell Biology in higher education;

✓   **Chapter II:** presents related works of Cell Biology and related areas, using the VR as a didactic resource;

✓   **Chapter III:** details the simulator construction process.

✓   **Chapter IV:** presents the analyses of the experimental research and the results obtained.

✓   **Chapter V:** encompasses the conclusion of this research and proposals for future work.

✓   **Appendix:** presents the documents developed in the course of this work.

Finally, the bibliographical references that subsidized the theoretical basis of the research are presented.

2. THEORETICAL FRAMEWORK

(Rubem Alves)

This chapter deals with the definition of VR, its history, its evolution, its concept, describes the basic characteristics of a virtual environment, presents the different types of VR currently used, with emphasis on their use in teaching. And it presents a bibliographical research, which demonstrates the complexity and abstraction of the teaching-learning of Cellular Biology, both from the point of view of the student and the teacher. Such research justifies the importance of a differentiated methodology to approach the contents of this discipline and, in this case, focusing on the use of VR.

2.1 Historical Context of RV

The first immersion experiment took place around 1956, with the creation of the filmmaker Morton Heilig: a video based simulator called Sensorama (figure 2). The "cinema of the future", as the machine was called, allowed the user a combination of three-dimensional vision, with stereo sound, vibrations, wind sensations and aromas in a simulated motorcycle ride through New York. The user felt "inside" the film, however, the invention was not a commercial success, but is considered the precursor of the user's immersion in a "synthetic environment" (RHEINGOLD, 1994).

Figure 2 - Virtual Reality Machine, Sensorama, created by Morton Heilig

Source: HEILIG, s/a.

18

RV in a very timid way had its beginning in the 1960s, with the invention of the RV linked to Ivan Sutherland's *SketchPad* interface. After this invention, the development of the VR stagnated for almost thirty years due to technological difficulties in the construction of elements necessary for the interface that could transport the user to the virtual environment with more comfort, such as helmets, gloves, stereoscopic glasses, 3D *mouse,* among others. Since the 1980's, the VR has evolved very fast due to the advance of microelectronics and the growth of industrial technologies and computer graphics (FIALHO, 2018).

In 1968, with the help of student Bob Sproull, Sutherland created the first VR system (figure 3), which was mounted on the head, affectionately referred to as the Sword of Damod, because, as the helmet was heavy, it was suspended from the ceiling, above the user's head. It was a machine designed to immerse the viewer in a simulated 3D environment (FIALHO, 2018).

Figure 3 - First Virtual Reality System, created by Sutherland

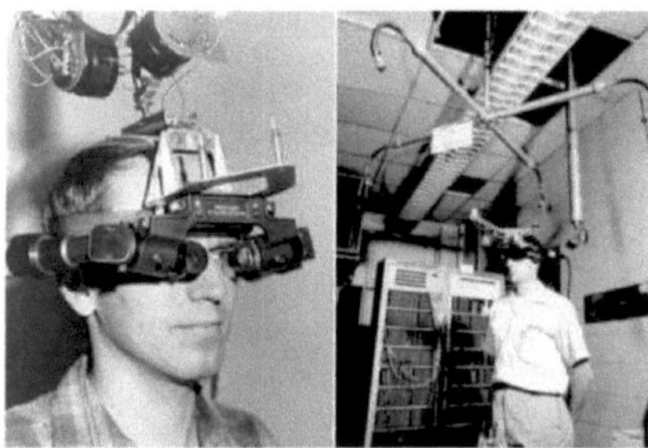

Source: GALDINO, 2019.

It is important to note that the VR and the various existing interfaces are not the exclusive result of Sutherland's research. Like all scientific research, this has also developed in parallel. There are records that, in 1958, the electronic equipment manufacturer Philco developed an interface that consisted of a pair of remotely controlled cameras and the prototype of a helmet with attached monitors that allowed the user to experience the feeling of being present within a certain environment. Later, this equipment was called *Head-Mounted Display* (HMD) (FIALHO, 2018; NETTO et al., 2002).

By the end of 1986, the *National Aeronautics and Space Administration* (NASA) team already had a virtual environment that allowed users to sort by voice command, manipulate virtual objects directly by hand movement. And most importantly, we saw the possibility of commercializing this technology, making acquisition and development costs more accessible. The awareness that NASA's ventures could generate marketable equipment initiated several RV researches around the world. From *software* companies to large computer corporations, they began to develop and sell products and services aimed at VR. In 1989, a *software* company, *Autodesk*, introduced the first RV system for personal computers (FIALHO, 2018).

Pioneer scientists in VR, such as Licklider, Engelbart and Sutherland, had their steps followed by others who independently or working together, built the basis of VR, preparing the ground for future research and applications, creating its prominent place in the area of Communication and Information and opening new horizons in the field of Education (CAMACHO, 1996).

2.2 Understanding RV

The VR consists of a combination of *software* and computers that allows the creation of a three-dimensional graphic environment, in which the user can move along the Cartesian axes x, y, z and also rotate around them, as shown in figure 4. This demonstrates the *software*'s ability to define the six degrees of freedom (6GDL), being three rotation movements and three linear or translation movements (forward and backward; up and down; left and right; upward and downward inclination; left and right angulation; left and right rotation) and the *hardware*'s ability to recognize them (CARDOSO; LAMOUNIER, 2006; LIMA, 1999).

Figure 4 - Navigation with 6 degrees of freedom (6GDL)

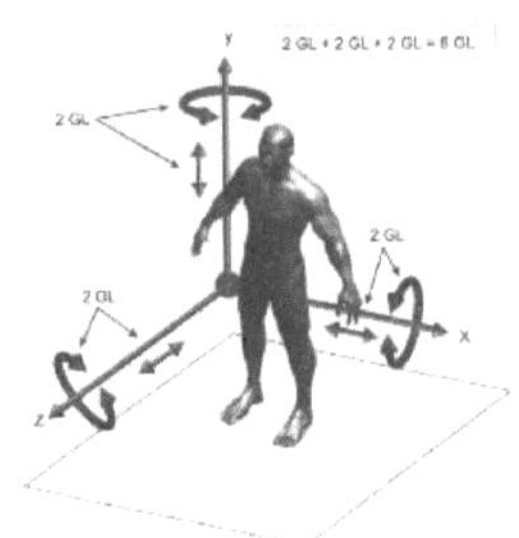

In essence, VR is a "mirror" of physical reality in which the individual exists in three dimensions, has the sensation of real time and the ability to interact with the world around him, and can even "touch" the objects that make up the virtual environment and make them respond or change, according to their actions (NETTO et al., 2002).

The goal of RV is to use interactive computer interfaces in a way that creates a sense of reality for the user. This sensation is achieved by means of seven basic characteristics described below and represented in figure 5 (MARTINS, 2000; VINCE, 2004).

✓ **Synthetic:** the environment is generated in real time and changes according to the interactivity of the user, that is, it is not a multimedia recording that is always repeated.

✓ **Three-dimensional:** it is necessary that the environment causes in the user the perception of a 3D environment, in which he can move and perceive the relations of his movements with the dimensions of width, height and length of the space in which he is interacting.

✓ **Multi-sensory:** the environment must be represented by more than one sensory mode (visual, sound, spatial, among others).

✓ **Interactive:** the user can modify the virtual environment in real time, because it reacts to the user's actions, through input devices such as sensors that are attached to gloves or boots.

✓ **Realistic:** the representation of the virtual environment and the real objects must be precise.

✓ **Immersive:** there must be an immersion of the user in the virtual environment, this characteristic is usually obtained by the use of multi-sensorial devices (helmet, glasses, gloves, controls, etc.) that capture the behavioral movements and react to them. This allows the sensation of presence within the environment.

✓ **With presence:** a subjective sense that gives the user the feeling of being physically inside the virtual environment.

Figure 5 - Basic characteristics of an RV environment

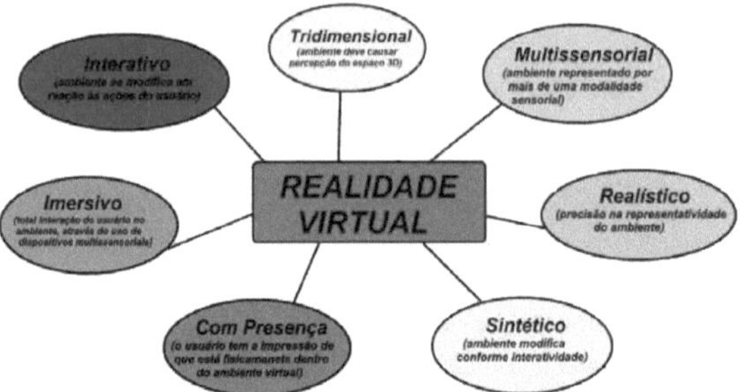

Source: author, 2019.

An important detail in VR visual interfaces that needs to be highlighted is stereoscopy, a natural phenomenon that occurs every time the individual looks at something: simultaneously two equal images are obtained from the scene, that is, one for each eye. This double exhibition occurs because in the real world when looking at an object, each eye sees a different image, that is, we see the same object, but in a different position.

This difference in image positioning is due to the pupil distance, which is the distance between the two eyes. When these two images reach the brain they are merged into a single image, decoded and superimposed giving the perception of depth, therefore the 3D image, which is how one sees in the real world. So, in VR *displays* that have as objective the immersion of the user, producing a sensation of being present in the observed environment, the screen is divided into two parts (figure 6), each one displaying the same image of the object for each eye, this causes a difference in positioning, in the same way that we see in the real world, soon added with hues of color, light, shadow and reflection, there will be the same three-dimensional perception of the real world (CADOZ, 1997; FIALHO, 2018; NETTO et al., 2002).

Figure 6 - Purely illustrative representation of the internal view of an RV visual interface

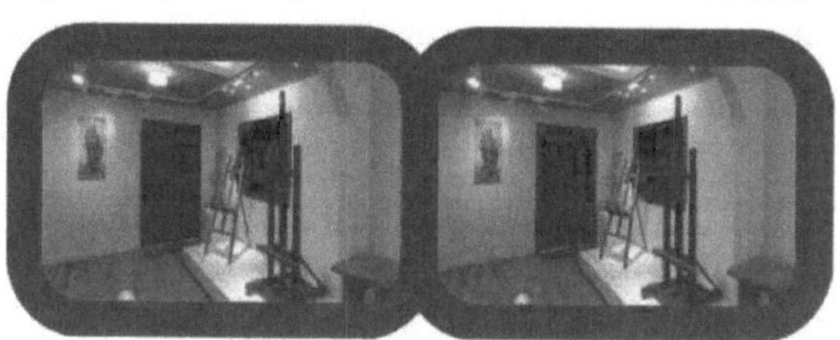

Source: FIALHO, 2018.

Depending on the user's sense of presence, the VR is classified as immersive and not immersive. In the immersive, there is total interaction of the user with the domain of the application due to the use of multi-sensorial devices (helmet, glasses, gloves, etc.) that capture movements and behaviors (input signals) and react to them (output signals), causing him/her a sensation of presence inside the virtual environment. In non-cured VR, the user is partially transported to the virtual world, through a window, which can be the monitor or a projection, but still feels predominantly in the real world (TORI; KIRNER, 2006). The Casa de Portinari museum is a good example of non-immersive VR[2]. Figure 7 exemplifies these two basic types of VR.

Figure 7 - Immersive and Non-Imersive PVR

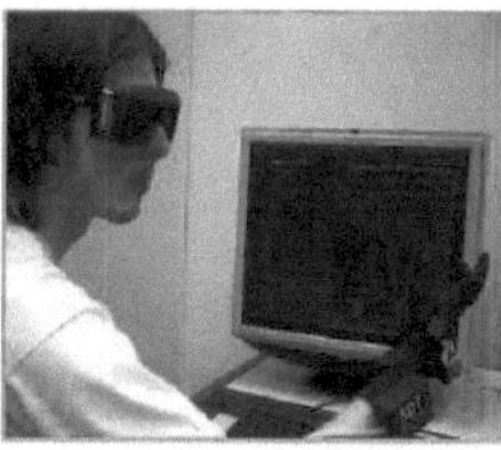

a) RV não imersiva com monitor.

b) RV imersiva com capacete HMD

2.3 Teaching of Cell Biology

Cell biology is the conceptual basis for biological knowledge, as it studies the structure, function and behavior of cells and serves as a prerequisite for other disciplines. Therefore, its contents have great relevance for the courses of Biological Sciences and Health courses, such as medicine, physiotherapy, pharmacy, biomedicine, nursing, among others, because it comprises the basic contents of the cell, which is the basis for the training of professionals in the area. It is a discipline covered in several nomenclatures (Cell Biology, Cell and Molecular Biology, Cytology, Cells and Molecules, General Biology).

During the entire learning process, the cell is viewed as something complex and abstract that is constructed in the mind of the students. And this occurs because it is a structure that can only be visualized by equipment that enables its study (PALMERO; MOREIRA, 1999).

One of the great difficulties in teaching cell biology is the visualization and understanding of cell structures and their functioning, since they are of microscopic sizes. The contents become abstract, built basically from microscopic and biochemical investigations, so it is difficult for students to understand, because in most cases the only resource available for their learning process is their imagination. And this makes it difficult to understand the importance of cell structures for organisms (ORLANDO et al., 2009).

The teaching of Cellular Biology is a complex content not only for students, but also for teachers, as the teaching-learning process is hampered by the poverty of didactic resources used by teachers and, consequently, the difficulty of creating mental models by students (OLIVEIRA et al., 2009).

a) Plasma Membrane Teaching Strategies

The plasma membrane (figure 8) is a dynamic and fluid structure that surrounds the cell defining its limits and maintaining the essential differences between the cytosol and the extracellular environment. It consists of a thin film of lipids and proteins that are bound mainly by non-covalent interactions. The lipid molecules are organized as a continuous double layer about 5 nm thick, which provides the basic structure of the membrane, acting as a relatively impermeable barrier to the passage of most water-soluble molecules. Membrane proteins usually pass through the lipid layer and measure almost all functions of the membrane, such as the transport of specific molecules through the membrane (ALBERTS et al., 2017).

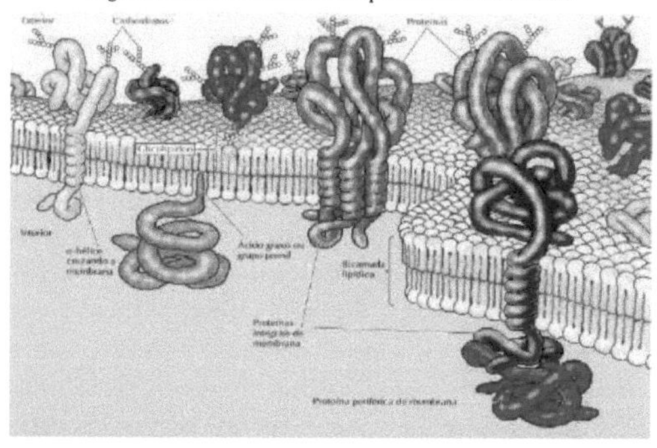

Source: COOPER; HAUSMAN, 2007.

The main teaching strategy adopted for the teaching of content on plasma membrane is the lecture, enhanced by digital tools such as *Power Point*, *Prezi*, videos or digital whiteboard resources. Looking for ways to propose the replacement of these passive methods and to improve the educational process regarding the study of Cell Biology, more specifically Plasma Membrane, innovative methodological alternatives are necessary to help the understanding, contextualization and construction of cellular knowledge of the students who study Cell Biology, whether they are students of Biological Sciences or of other areas.

It is important to use methodologies that allow the student to handle instruments and devices related to cells, instead of having a teaching based only on the use of textbooks and analogies that often cause confusion and generalizations of scientific concepts (REIMEIER; GROPENGIEBER, 2008).

After all, the teaching of Cell Biology uses many scientific concepts, terms and methodologies, taught in a way that does not arouse any motivation in the student, so they do not feel attracted by these contents (ZUANON et al., 2010).

However, the use of animations as a tool in the teaching of Cell Biology facilitates the teaching-learning process, because the use of animations guides the student in the abstraction of transformations of an image over time and also contributes to saving time, because it is much easier to learn when observing a process than when reading its explanation (MENDES, 2010).

2.4 Virtual Reality in Education

Teaching needs to be constructive, so the transmission of information is not the most appropriate means for the teaching-learning process. The student needs to build his own knowledge and the teacher needs to provide the necessary tools to make this possible. Offering dynamic and exploratory teaching that provides students with the opportunities they need to develop their skills and competences as well as overcome their difficulties (BEHRENS, 2013).

Due to its specificities and attributes, RV has great educational potential and constitutes a versatile and effective learning environment in which students can learn from their personal exploitation, thus building their own knowledge. The possibility of creating virtual worlds in which users are able to move, see, hear and manipulate objects as they do in the physical world represents an important teaching resource, since these simulated realities enable students to discover, in an active and playful way, the knowledge that was previously transmitted only by the teacher (BRAGA, 2001).

Immersion in the virtual world is an interesting factor of VR in education and teaching, because the student, entering the simulation, discovers by himself and learns by building his knowledge through experienced VR, based on the sensations he perceives. In this way, learning becomes more alive, richer, more varied and at the same time more lasting, because as it constitutes a personal experience, it will hardly be forgotten (CARVALHO, 2002).

The digital age brings a new way of categorizing knowledge, without having to discard the lectures. Electronic resources can be used as tools to build more meaningful methodological processes that facilitate learning. The VR in the digital era can be used as a learning resource, these innovations already exist, it is now up to educators to appropriate this technology and create projects that lead their students to "travel" in a virtual environment and benefit from this technology of VR, which lead the student to "learn to learn" with interest, creativity and autonomy (BEHRENS, 2013).

3. RELATED WORKS

This section brings a bibliographic survey, carried out in scientific repositories, of works that address the same theme of the researcher.

All research implies a survey of data from various sources, whatever the method or technique employed. The type of research used for the data collection of this document was the bibliographic research or research of secondary sources, such as consultations to bibliographies already published in the form of books, magazines, single publications and the written press. The main purpose of the bibliographic research is to "put the researcher in direct contact with everything that has been written about a certain subject" (MARCONI; LAKATOS, 2001).

> The bibliographic research is that which is carried out from the available register, resulting from previous researches, in printed documents such as books, articles, theses, etc. It uses data or theoretical categories already worked by other researchers and duly registered. The texts become sources of the subjects to be researched. The researcher works on the basis of the contributions of the authors of the analytical studies contained in the texts. (SEVERINO, 2007, p. 122).

The bibliographic research sought to compile academic-scientific productions, in Portuguese and English, that address plasma membrane and/or transport of molecules through the plasma membrane, using VR technology. This research was carried out through the *internet*, through the scientific repositories *PubMed, Science Direct* and *Scielo*, in the months of March and April 2018, and as the quantity of correlated works found was very scarce, the author judged it necessary to redo the research, in the months of September and October 2019.

The objective of this research was to identify if there is any teaching tool developed in VR similar to the one being proposed in this work. And if there is something similar, to analyze if the methods used are similar to what is being proposed.

The searches were made using the keywords: virtual reality biology/virtual reality *biology; virtual reality* sciences/virtual reality *science; virtual reality in* teaching/virtual *reality in teaching;* the *use of virtual reality in teaching practices;* use of virtual reality in the teaching *practices of the biological sciences courses; virtual reality in cell biology/virtual reality in cell biology; virtual reality in plasma membrane teaching.* These keywords have been combined with each other in order to obtain a greater amount of results and at the same time filter out unwanted results.

Chart 1 presents the keywords and scientific repositories used in the search for related academic papers and the number of papers found for each keyword.

Chart 1 - Keywords used in the search X number of works found

NUMBER OF JOBS FOUND			
KEY WORDS	**PUBMED**	**SCIENCE DIRECT**	**SCIELO**
Virtual reality biology	0	18	0
Virtual reality biology	219	7.325	0
Virtual reality sciences	0	196	13
Virtual reality science	1.694	43.414	28
Virtual reality in education	0	78	20
Virtual reality in teaching	2.951	12.020	40
The use of virtual reality in teaching practices	0	49	2
The use of virtual reality in teaching practices	36	9.564	3
Use of virtual reality in teaching practices of biological science courses	0	2	0
Use of virtual reality in the teaching practices of the biological sciences courses	0	1.160	0
Virtual Reality in Cell Biology	0	3	0
Virtual reality in cell biology	39	4.180	0
Virtual reality in plasma membrane teaching	0	0	0
Virtual reality in plasma membrane teaching	1	270	0

Source: author, 2019.

Despite the high results obtained during the research, mainly in the *Science Direct* scientific repository, the works were not related to the teaching of Cell Biology using VR. The analysis of the results obtained was initially by reading the titles, as the result of some repositories exceeded 1,000 works, the researcher established that the titles of the first 100 works should be read, because, as they progressed, the titles of the works were more distant from the term researched. Many papers appeared as a result of more than one keyword.

After reading the titles, 31 articles were selected, 12 from the *Pubmed* repository, 14 from *Science Direct* and 5 from *Scielo*, which were initially examined by reading the abstracts. In *Pubmed*, the works were found with the keywords *virtual reality biology* (eight works) and *virtual reality in cell biology* (four works); in *Science Direct*, the results were obtained with the keywords virtual reality *biology* (ten works), virtual reality in *teaching* (two works), virtual reality in *plasma membrane teaching* (two works); and in *Scielo*, two with the keywords *virtual reality biology, and* three with *virtual reality science*.

Then a new screening was made of the articles that were possibly related to the content of the researcher's work, the abstracts were analyzed and those that contained VR productions in the teaching of Cell Biology were selected. Although none of the articles found was directly associated with the teaching of plasma membrane and VR technology, the works that came closest to the product proposal of this dissertation are listed in the next section.

3.1 Articles Found During the Bibliographical Research Relating Cell Biology to VR

After reading the abstracts, 17 articles were selected and read in full. In 12 articles, there were reports of VR experiences associated with the teaching of Cellular and Molecular Biology. These experiences will be reported below.

a) Exploring Molecular Structure by Virtual Reality (HARTSHORN et al., 1995).

The authors present the development of VR tools for visualization of DNA molecular structures, through modeling that used to be static and now can be rotated in real time on a two-dimensional graphic screen. Their model did not yet represent a truly three-dimensional modeling. The DNA molecule is complex, so it is important that the student visualize and interact with these structures (figure 9A), the modeling of DNA is visualized through the VR glasses (figure 9B).

Figure 9 - Representation of DNA and RV glasses

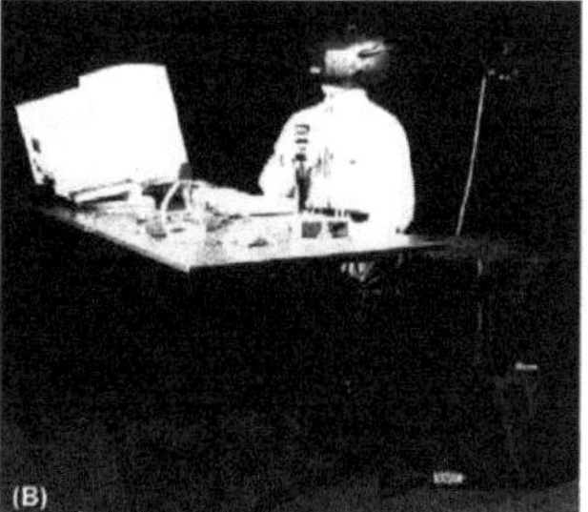

Source: HARTSHORN et al., 1995.

b) The Gentechnique Project: Developing an open Environment for Learning Molecular Genetics (CALZA; MEADE, 1998).

29

The authors emphasize the importance of differentiated teaching methodologies, especially for such abstract contents as the study of molecular genetics and biotechnology that requires students' understanding of Cellular and Molecular Biology, such as DNA replication and protein synthesis. They describe how Washington State University, together with the Faculties of Science and Agriculture, used the *GenTechnique* project, which aims to build a learning environment for molecular genetics, in order to improve the academic performance of the students. The environment features animations, hypertext and interactive exercises.

This project, through a *Web-based* learning module on DNA replication, promotes student motivation and allows them to build their knowledge. For that, the themes are developed in two ways: textual narrative and graphic visualization, which include interaction between molecules, enzymes and cell structures. The authors state that the use of *GenTechnique* facilitates the learning process.

c) A Virtual Environment for Steered Molecular Dynamics (PRINS et al., 1999).

In this research, the authors describe an application of the *Steered Molecular Dynamics* (SMD) system to extract small proteins in an immersive 3D virtual environment, where it is possible to visualize and interact with three-dimensional structures of proteins. The structure of the molecule can be handled, moved and observed from different angles, removing or adding bonds.

The authors developed an integrated virtual environment system for SMD using the *Protein Interactive Theater* (PIT) *software*, a dual screen stereo display system for two operators simultaneously. Each user uses glasses and controllers that display an image with sound and motion and provide them with 6GDL and manipulations (figure 10).

Source: PRINS et al., 1999.

The authors believe that the SMD system is a valuable tool for biophysicists in the execution of a computational analog of such experiments, because it improves the understanding about molecular forces and their reactions.

    d) Simulated Site-directed Mutations in a Virtual Reality Environment as a Powerful aid for Teaching the Three-dimensional Structure of Proteins (SOTRES et al., 2009).

The authors state that molecular visualization of the consequences of a mutation can assist undergraduate and graduate students in teaching the relationship between the structure and activity of a protein. Then, they created a database with models of subtle (alanine) or aggressive (yeast pyrophosphatase homodymer) mutations, displayed in *Visual Molecular Dynamics* (VMD) a computer program for molecular modeling, which was visualized by students at IXTLI, an immersive VR observatory of the National Autonomous University of Mexico (UNAM).

A user interface has been programmed to create the illusion of mutations in real time with a mouse click. During the immersion, students are guided until they learn how to explore the *software*. Student interaction varied substantially, but most were highly motivated and seemed to have benefited from using VR.

The user interface includes the 3D protein structure image generated in an OpenGL VMD window and a user control *widget* written directly in Tcl/Tk 8.4 linked to the VMD

31

commands, which allow you to rotate on the X, Y and Z axes, control the zoom and select what you want to view.

The authors validated the product with 16 undergraduates from the Faculty of Chemistry at UNAM who had already studied Cellular Biochemistry. The validation started with a theoretical approach about the structure of proteins. Before starting the VR session, a brief description of the enzyme under study was made and it was explained how the *software* works. The students were separated into four teams, and each had approximately ten minutes to respond to an activity using the *software*.

At the end of the activity, they were invited to give their personal impressions about using the *software* as a learning tool, evaluating their own performance. In general, they evaluated the VR environment as stimulating and complete, scoring that the experience allowed them to feel able to "travel within the protein" and visualize factors that they could not see in the two-dimensional representations of proteins. Interestingly, in their comments, everyone referred to an improved understanding of the interactions between amino acids and their role in structural changes in proteins.

In addition to the students, the authors invited professors from the Faculty of Chemistry at UNAM to use the product. All the invited professors taught biochemistry at undergraduate and/or graduate level. Surprisingly, and unlike the students, most of them were reluctant to take the controls. Only 2 of the 20 professors present at the validation of the product presented themselves to manipulate the images personally. At the end of the session, they evaluated the *software* as a useful teaching aid and suggested some changes to increase the complexity of the interface and its resources.

 e) Journey to the Centre of the Cell: Virtual Reality Immersion into Scientific Data (JOHNSTON et al., 2017).

In this article, the authors highlight how they used state-of-the-art imaging techniques to create a 3D virtual model of a cell from image data obtained by the block scan electron microscope (SBEM).

They created a breast cancer cell model, the modeling was done in *Autodesk Maya*[3] and then exported to *Unity software*. The model created allows scientists, students and members of the public to explore and interact with a "real" cell.

---

[3] Computer animation *software*, modeling, simulation and 3D rendering.

Initial tests of this immersive environment indicated a significant improvement in students' understanding of cellular processes. The authors speculate that VR can become a new tool for researchers studying cellular architecture and processes, filling VR models with molecular data.

To facilitate the user's understanding at the time of immersion, they created two differentiated VR environments. One they named "*paddock*", which is open and allows exploring the surface of the cell, and the other called "cathedral", where the interior of the cell can be explored.

The product was tested with graduates of the Bachelor of Pharmacy course. First, they had a theory class on the subject and then were invited to explore the VR cell, 48 students chose not to explore the cell and each of the 15 students who volunteered to explore the cell had 15 minutes to do so. Then, the 63 students who were participating in the research were submitted to a test in which they had to explain the mechanism by which a particle could be internalized in a cell, and what the likely destination of the nanoparticle would be.

Analyzing the result, the authors state that the students who navigated in the VR cell performed 5% better in the question of Cell Biology than in the rest of the exam. In comparison, the students who did not try the VR performed 35% worse on the question of Cell Biology. These preliminary results suggest that VR can have a positive and significant impact on students' understanding and learning.

f) From Atoms to Cells: Using Mesoscale Landscapes to Construct Visual Narratives (GOODSELL et al., 2018).

The authors highlight the advances in experimental and computational methods that allow the materialization of abstract contents of Molecular and Cellular Biology. The aim of the authors is to create accurate representations of the molecular structure of cells, integrating data from Structural Biology, Biochemistry and Cell Biology.

The authors have used several *software* that present modeling of cell structures and also allow new structures to be created, such as *MolecularMaya*[4], *BioBlender*[5], *GraphiteLifeExplorer*[6] and *CellPAINT*[7].

The authors state that 3D illustrations are useful in the teaching-learning process because they allow a better exploration of cellular structures by students. They also point out that students often memorize the content without understanding it, so they quickly forget them. In order to remedy this difficulty, the authors used the pages to color in grayscale and the *flash* memory cards that accompany a panorama cellular landscape, both created by the BioMolecular Modeling Center.

g) Virtual Reality: Beyond Visualization (BEHEIRY et al., 2019)

The authors' proposal is, through VR, to use single molecule localization microscopy, which is a key methodology to explore cellular function, but difficult to be located in a 2D image, therefore, VR can be a fundamental tool in this task.

The authors represented, in the virtual environment, neuron images of the olfactory bulb of mice for users to develop segmentation activities. The controllers allow the mapping of the cell interior and the measurement of axon segments (figure 11).

---

[4] A free *plug-in* for *Autodesk Maya* that allows users to import, model, and animate molecular structures (https://clarafi.com/tools/mmaya/).

[5] An implementation of Blender, with BioBlender, users can manipulate proteins in 3D space, displaying their surface in a photorealistic manner and elaborate protein movements based on known conformations (http://www.bioblender.org/).

[6] A modeling tool that allows you to create 3D molecular assemblies of proteins and DNA from Protein Database (PDB) files. Atomic DNA can be modeled from scratch or reconstructed from the simulation (http://www.lifeexplorer.info/).

[7] Use the digital painting approach (http://cellpaint.scripps.edu).

Figure 11 - Use of RV in a segmentation task

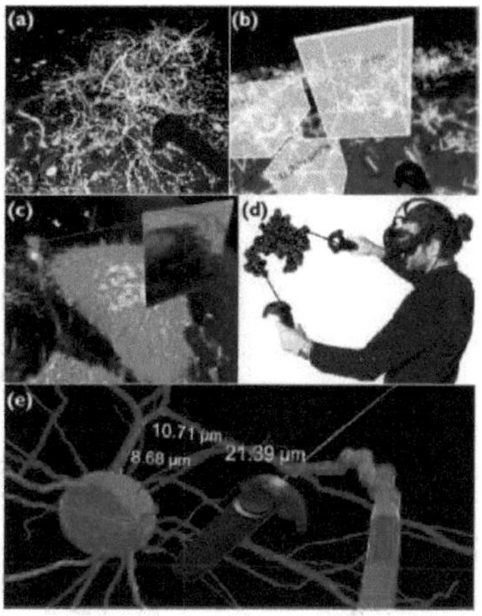

The authors believe that VR will have an important impact on Cell Biology activities, as 3D visualization provides an intuitive visual understanding of volumetric data, which can promote rapid transmission and training of knowledge in the laboratory.

h) Exploration of an Interactive "Virtual and Actual Combined" Teaching mode in Medical Developmental Biology (XU et al., 2018).

In this article, the authors analyzed the efficiency of VR in teaching. The research was carried out at the Medical University of China (CMU), with 453 medical students, and the experiment was carried out in the discipline Developmental Biology and Regenerative Medicine, which has many contents related to Molecular Biochemistry, considered by the professors as a very difficult discipline for students, because the biochemical molecular techniques can only be visualized at a microscopic level, that is, it has an abstract nature. The use of VR in this discipline can assist students in understanding the theory, as well as providing scientific and laboratory practices.

The Virtual Simulation Lab (VSL) was built in a highly realistic way, so it relates very well to the actual practice of the lab; and as a result, it simulates interesting and motivating learning and saves teachers time and effort in preparing the practical classes.

By using the CMU VSL platform, students can easily connect to open learning resources using the student ID, where they have access to educational video and various related interactive hands-on exercises. In the discipline of Biology Development and Regenerative Medicine, there are currently several hands-on classes aimed at training students to deal with various techniques involving biochemical and molecular principles.

The authors report that after completion of all practice sessions, teachers found that the learning outcomes were fully achieved. In addition, students' motivation to learn had been increased using the VSL system. Collecting the *"e-learning"* data, which includes the time students stay *online* and their final practical scores, they found that 96.9% of the students worked more than two hours *online* at times contrary to what they attended university, and 15% of them worked more than five hours. The authors see these results as encouraging, as they indicate that students seem to be very interested in learning using VR.

The results of the research show that students have significantly improved their ability to analyze data and understand scientific results, which is highly necessary in their future jobs, as well as their procedural knowledge, motivation and interest in the topic. And in the questionnaire given to them to get *feedback* on the teaching methods used in the VSL platform, they quantified that 100% of the students considered the VSL platform useful for learning; 95% of the students mentioned that the use of VL facilitated the understanding of topics, which was impossible with the theoretical class alone.

i) A Practical Guide to Developing Virtual and Augmented Reality Exercises for Teaching Structural Biology (GARCIA-BONETE et al., 2019).

The product presented in this article makes use of VR and augmented reality (AR) in Structural Molecular Biology. The protein molecule can be rotated by clicking the button and sliding the *mouse in* its direction. The *Google Daydream Vista* system was used along with a *Samsung Galaxy S8 smartphone* as the RV platform.

To develop the product, *Sketchfab* was used, a platform for publishing and sharing 3D structures. The 3D models loaded in *Sketchfab* can be private or public. The Web RV viewer for *Sketchfab* is free. The *Android Google Chrome* was used for visualization. *Oculus Rift* and *HTC Vive* can also view *Sketchfab* through *Google Chrome/Chromium* or *Firefox, web* browsers. *Sketchfab* applications for *Android* and *iOS* can be downloaded for free from the *Play Store* and *App Store*, respectively.

The virtual model was made available in a room and test participants could walk around and watch through the *smartphone* screen from different angles[8]. Each learning task was entirely based on a multi-platform 3D model uploaded to the *Sketchfab* website, available under a *Creative Commons* license.

The virtual protein model was developed because molecular visualization aims to show the arrangement of atoms so that the observer can perceive and interpret, and often the atoms are visualized as balls with colors corresponding to the type of element. But, for protein visualization, this representation is not ideal because they are made up of thousands of atoms. So the main protein chain is usually replaced by a cartoon model that follows the shape of the carbons in the protein chain, that is, the style of the design differs according to the secondary structural elements that make up the protein.

The VR/RA model was based on the atomic structure of *Blastochloris viridis* bacteriumferritin, represented by a cartoon. Each monomer was colored differently to highlight the quaternary structure of bacterioferritin. In one of the monomers, the site of the catalytic ferroxidase was developed so that they could visualize the ball and stick model, two iron ions and atoms that belong to the coordinating amino acid residues. Similarly, one of the heme-b molecules is visualized by a ball and stick representation.

The VRML model was uploaded to the *Sketchfab* repository, after the *upload*, the model received a name, a brief description and keywords, being assigned to the Science and Technology category. In the VR, the model was placed in front of the viewer and the initial scale of the model was set to approximately twice the height of the viewer, so that the eye level coincides with the origin standing inside the virtual molecule. A sound file containing a two-minute narration about the symmetry and different characteristics of bacterioferritin and its biological role was associated.

The main contents addressed were bacteriopherritin structure-function, ferroxidase activity in the transport and storage of iron ions. The authors emphasize that it is important for the student to visualize the active site where the enzymatic catalytic reaction occurs and understand its relationship with the protein support.

---

[8] The bacterioferritin scene and the voice associated with the recording were shared through the link https://sketchfab.com/3d-models/bacterioferritin-biological-unit-b8b4517e54884f3d98b02e538d37d4d4 and are available for viewing, downloading and modification under Creative Commons Attribution 4.0 International (CC BY 4.0)

The product was tested at the Department of Chemistry and Molecular Biology, University of Göteborg, and 13 students participated in the voluntary test, being 3 postdoctoral students, 6 doctoral students, 3 master's students and 1 graduate student. This variety is due to the fact that biochemistry and structural biology are multidisciplinary environments, where chemists, biologists, physicists and engineers work together.

The test was done individually and each participant performed the tasks of VR and AR, under the supervision of a PhD student who briefly explained how the equipment was used.

After analysis of the questionnaire answered by the participants, the authors concluded that, overall, the experience was positive. Most users evaluated that the narrative in the audio was not useful, because they could not really concentrate on what they heard, because their attention was focused on what they were visualizing. Regarding the visualization, they were fascinated and there was a report about the uniqueness of the experience, because they did not think that the VR/RA was substantially different from the visualization of the structure on a computer screen, besides, they liked the possibility of rotating and repositioning the structure.

As for the question "Could you suggest another area where RV/RA can improve higher education? ", suggested several contents and areas, such as physiological visualization, surgical processes, visualization of art objects that are normally not within public reach. The diversity of responses reflects the interdisciplinarity that VR/RA can provide.

j) Virtual Laboratory Simulation in the Education of Laboratory Technicians-motivation and Study Intensity (MAY; VRIES, 2019).

In this article, the authors present an evaluation of the educational use of virtual laboratories in the Graduate Program in Chemical and Biotechnical Science of the University *College Copenhagen*, Denmark. The research was conducted between 2014 and 2015, with 78 students. The objective was to test how laboratory simulations, more specifically Labster [9], could be applied to a teaching oriented educational practice.

The virtual laboratory resembles a "real" laboratory (figure 12). With the *mouse*, the student moves around the virtual laboratory and performs several laboratory tests. A virtual laboratory assistant guides him during the activities. The virtual experiments are interrupted by multiple choice questions, which need to be answered correctly before proceeding (figure 12D), generating points. The teacher can follow the individual answers of the students and also check if the students have completed the simulation.

Figure 12 - Illustration of a Labster Virtual Lab

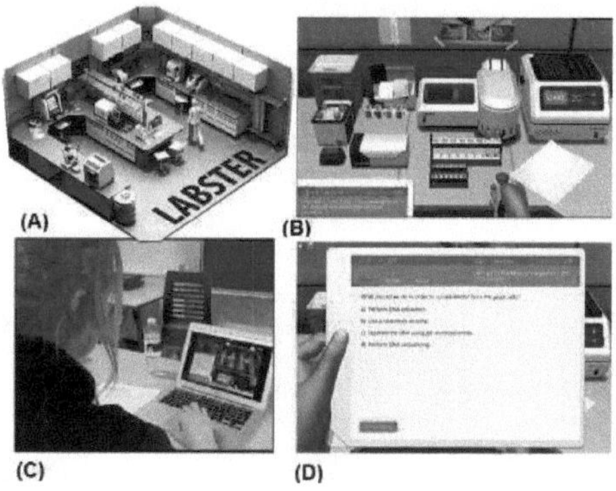

Source: MAY; VRIES, 2019.

---

9 Labster is a company dedicated to the development of advanced, fully interactive laboratory simulations based on mathematical algorithms that support open investigations.

For this research, NGS (*Next Generation Sequencing*) and Molecular Cloning laboratories were used. In the NGS laboratory, the simulation is short and, to conclude, the student uses about 20 to 40 minutes. The simulation is around an extinct pallaeo-skeleton, where the student obtains a bone sample from it, to extract the DNA and analyze its sequence of nitrogenous bases. In the Molecular Cloning simulation, the process is more complex and long, taking approximately 2 hours. The simulation is built around a "story" about a researcher seeking to test a protein (RAD52) for its hypothetical function in DNA repair. The simulation starts with an animation film where the researcher receives inspiration on how to assemble the experiment using the green fluorescent protein (GFP) from jellyfish. During the virtual experiment, students work with different molecular techniques, including DNA extraction and restriction enzyme. They perform actual analysis to determine whether RAD52 plays an important role in DNA repair.

In general, most students positively evaluated the use of the laboratory simulator, highlighting it as a useful complement to traditional teaching and describing that the laboratories helped in visualizing practical procedures as well as the chemical and molecular reactions that occurred during the experiments. In addition, they said that virtual simulation helped in understanding the theory as well as the elements of laboratory work practice, which provided them with better learning.

The authors attributed that there are two aspects of how the virtual laboratory helps learning: 1) The animation of molecular processes helps the student to understand the theory; 2) The virtual laboratory with visualization of equipment and techniques helps the students to perform practical procedures. They concluded that the virtual laboratory has proven to be a very promising teaching tool, as it can be used by students of different teaching levels and allows a constant interaction of the student, making him/her active throughout the simulation.

    k) Equivalence of Using a Desktop Virtual Reality Science Simulation at Home and in Class (MAKRANSKY et al., 2019)

The authors of this article researched the impacts of using a VR laboratory for learning. The research was carried out with a sample of 112 biology undergraduates from the University of Glasgow, Scotland. This sample was divided into two groups, one of 62 students who used the virtual laboratory simulation at home in their own time, and the other group of 50 students made use of the same simulation, but in the classroom, under the supervision of the professor.

The interactive virtual simulation used in the research was entitled *"Bacterial Isolation"*[10], a laboratory of a catalogue that is available at Labster, developed to facilitate the learning of the main concepts and techniques in microbiology.

Specifically, the concepts addressed were: the importance of bacterial growth for the investigation of pathologies; the need to work under aseptic conditions; colony concept; scratching techniques; function and importance of selective culture media. The simulation allowed the user to work through the procedures in a virtual laboratory using and interacting with the relevant laboratory equipment and the essential content is taught through a question-based learning approach. Students are guided through the simulation by a pedagogical agent who provides the specific instructions, which assists in the students' progression during the simulation.

The virtual classroom simulation begins with a student being presented through a brief introduction to the story behind the laboratory setting and then describes the student's task in solving experimental research. The student needs to investigate an outbreak of bacterial food poisoning, and has to isolate an antibiotic-resistant strain of the bacterium from samples taken from a poultry farm, which is a suspected source of infection. After the introduction of the principles of selective and differential culture media in microbiology, students have several opportunities for practice by listing bacteria on agar plates, incubating them properly, and isolating cultures from colonies free of any contamination.

This simulation allows five different forms of interactivity that are commonly used in multimedia learning environments, including: dialogue, control, manipulate, search and navigate. The dialogue is achieved through *online* interaction with the virtual lab instructor and through the optional selection of additional information via *wikilinks*. Students are also able to control the pace of the bacterial isolation simulation, deciding when to proceed with experiments, choosing whether to take more readings, when asked to answer multiple-choice questions, and controlling the number of times they practice removing bacteria from agar plates. This includes selecting innocuous culture media and appropriate controls used in the experiment.

The student needs to find the right tools and prepare them adequately to work with the sterile technique required by the protocol and finally has to determine the parameters for successful incubation and growth of bacteria. The key to the simulation is the student's ability to use a sterile nicronomy *loop* to scratch bacteria on an agar plate in order to grow isolated,

---

[10] Which can be accessed through this link: https://www.youtube.com/watch?v=zZYUob44efE.

contamination-free bacterial colonies. The virtual laboratory also requires the student's choice for the safe disposal of contaminated laboratory materials, and offers the opportunity to become involved in the search for information by providing written material, supporting the concepts, techniques and materials. Finally, students also receive interaction with navigation because they are in a virtual laboratory, where they are able to determine the content of learning episodes by selecting equipment from various available sources and decide what to do in the virtual laboratory.

The classroom group had the simulation scheduled on their schedule, allowing them one hour in a computer lab on *campus*, which could accommodate up to 40 students at a time, and the same assistant (a PhD student in the last year) was present in each session. On average, the students used 59.21 minutes in the simulation. Most of the students in the group who used the simulation outside the University chose to do the simulation at home (52 students), only 10 students reported doing the simulation elsewhere, as a library. The students in this group spent on average 61 minutes using the simulation.

After the end of the simulation, the students of both groups were submitted to an evaluation and, according to the results presented by the authors, the groups did not differ significantly in relation to the learning. The authors concluded that the virtual context of a carefully designed scientific VR simulation can be an excellent pedagogical tool, as students are immersed in this virtual context and can experience a sense of presence, regardless of the environment in which they are physically present.

l) A Virtual Tour of the Cell: Impact of Virtual Reality on Student Learning and Engagement in the Stem Classroom (BENNET; SAUNDERS, 2019).

In this article, there is an account of the virtual experience of 62 undergraduate students in the cellular biology class at Otterbein University. This experience was through *Journey Inside a Cell*, created by *The Body VR* using [11]*HTC Vive*.

There were two HMDs available with the "Travel within a cell" module, which provides a virtual guided and immersive tour lasting approximately 12 minutes. The student can interact with the cell and its cytoplasmic organelles, manipulating them through the controller, in addition to "shooting" antibodies in an approaching "viral attack". At the end of the virtual experiment, a challenge was proposed, in which the student needed to classify the cells. Working in pairs, they combined the parts of printed VR cells with the correct labels.

After the virtual experience, students participated in a survey that aimed to evaluate the impact that total immersion VR in the cell brought to the learning of Cell Biology. Analyzing the results of the research, the authors found that 93% of the students stated that VR facilitated their learning and the most found justifications for this were grouped into categories: increased interest; better understanding and new perspectives.

3.2 Comparison of the Works

Of the 12 papers presented above, all those that presented validation were positively evaluated. And the validations were very similar with the testing done on the product of this dissertation, being first of all volunteers from the undergraduate and/or higher education teachers invited to use the simulator and evaluate it through a questionnaire.

In all the works raised it became evident how important the modeling of abstract contents of Cellular Biology is for the teaching process, and how this reflects positively on learning, which reinforces the importance of this study.

Not all VR environments present the seven characteristics, highlighted in figure 6 and presented in chapter I of this research, besides differentiating according to the types: immersion (immersive and non-cure); scenario (real or fictional); interaction; consoles used; among others.

Considering some characteristics of virtual environments, a comparative table (Table 2) was made with the 12 papers presented above and these were numbered as follows:

I. *Exploring molecular structure by virtual reality;*

II. *The gentechnique project: developing an open environment for learning molecular genetics;*

III. *A virtual environment for steered molecular dynamics;*

IV. *Simulated site-directed mutations in a virtual reality environment as a powerful aid for teaching the three-dimensional structure of proteins;*

V. *Journey to the centre of the cell: virtual reality immersion into scientific data;*

VI. *From atoms to cells: using mesoscale landscapes to construct visual narratives;*

VII. *Virtual reality: beyond visualization;*

VIII. *Exploration of an interactive "virtual and actual combined" teaching mode in medical developmental biology;*

IX. *A practical guide to developing virtual and augmented reality exercises for teaching structural biology;*

X. *Virtual laboratory simulation in the education of laboratory technicians-motivation and study intensity;*

XI. *Equivalence of using a desktop virtual reality science simulation at home and in class;*

XII. *A virtual tour of the cell: impact of virtual reality on student learning and engagement in the stem classroom.*

Table 2 - Related works and the characteristics of the VR

WORKS CORRELATES	SYNTH ETIC	3D	MULTISENS ORY	INTERA- TIVO	ITALIES	IMERSI- VA	WITH PRESENCE	HOW INTERACTION IS MADE	AVAILI- BILIZATION	CATE- GORIA
I	*		*	*	*	*	*	It uses immersion glasses, but in the research it does not specify the model.	No information if available	Education al
II		*	*	*				Visualizes on the computer screen and the interaction is done through the *mouse*	Free	Education al
III	*	*	*	*	*	*	*	PIT display system	No information if available	Search
IV	*	*	*	*	*	*	*	In sessions at the IXTLI, the UNAM Virtual Reality Observatory.	In the research, there is no information if it was made available	Education al
V	*	*	*	*	*	*	*	*Rift and HTC VIVE Eyeglasses*	Authors do not inform if the product has been made available	Education al

WORKS CORRELATES	SYNTH ETIC	3D	MULTISENS ORY	INTERA-TIVO	ITALIES	IMERSI-VA	WITH PRESENCE	HOW INTERACTION IS MADE	AVAILI-BILIZATION	CATE-GORIA
VI	*	*	*	*	*	*	*	They show that there is immersion and interaction, but do not inform the console used for this.	Authors do not inform if the product has been made available	Education al
VII	*	*	*	*	*	*	*	*HTC VIVE Eyeglasses*	Authors do not inform if the product has been made available	Education al
VIII	*	*	*	*	*			Through CMU VSL platform	Free	Education al
IX	*	*	*	*	*	*	*	*Android Google Chrome Oculus Rift HTC VIVE*	Free	Education al
X	*	*	*	*	*			Through Labster	Paid	Education al
XI	*	*	*	*	*			Through Labster	Paid	Education al

WORKS CORRELATES	SYNTH ETIC	3D	MULTISENS ORY	INTERA- TIVO	ITALIES	IMERSI- VA	WITH PRESENCE	HOW INTERACTION IS MADE	AVAILI- BILIZATION	CATE- GORIA
XII	*	*	*	*	*	*	*	*HTC VIVE* *Oculus*	Free	Education al

Source: author, 2019.

Seven papers present all the characteristics of a virtual environment described by Martins (2000) and Vince (2004), explained in chapter I of this dissertation. Among the 12 papers presented, only one was not developed for education. And regarding the functionalities of the resource, four use the same *headset* of this research, the *HTC Vive Oculus*.

With the completion of the bibliographic research, it is noted that technological advances and new teaching methodologies and techniques in the area of education allow a new vision for teaching focused on the study of cells, a content of extreme relevance to understand the areas of Biological Sciences. The works presented allow the teacher to make use of innovative and diversified methodologies that allow a better understanding of the content, instead of being restricted to expository classes and passive methodologies, with the use only of *slides*, videos and construction of mockups with three-dimensional models in which the students only receive the information, but are not able to interact with them.

The work that most closely resembles the product of this dissertation is the experience of *Journey Inside a Cell*, created by *The Body VR* and described in the work of Bennett and Saunders (2019), because both allow the user to interact with the cellular components present in the virtual environment and also allow the simulation of processes that occur inside the cell.

Thus, this work, seeking to collaborate in the cognitive process of the apprentice, proposes a model of the plasmatic membrane in VR, which provides an exploratory education and offers the apprentice the opportunity to better understand its object of study.

4. PRODUCT CONSTRUCTION PROCESS

This section contains the entire process of building the simulator in a virtual reality environment for teaching the plasma membrane. Describing the entire modeling process carried out by the researcher and explaining the process of exporting to the *Unity* application, carried out by the LAAI team of programmers.

4.1 Modeling of Structures

For the construction of the plasmatic membrane model in VR, it was decided to use the Blender 3D modeling *software* version 2.78, distributed under the GNU-GPL (*General Public Licence*) license, which allows the construction and rendering of three-dimensional objects. This *software* can be used to create three-dimensional space views, still images, as well as high-quality videos (BRITO, 2011).

The plasma membrane has a lipoprotein constitution and the shaped structures were lipids: cholesterol and five types of phospholipids (sphingomyelin, phosphatidylcholine, phosphatidylethanolamine, phosphatidylinositol and phosphatidylserine); eight types of proteins (single α helix transmembrane protein; multiple α helix transmembrane protein; single leaf transmembrane protein β; protein totally exposed to the external surface of the cell; protein located entirely in the cytosol and attached to the cytosolic monolayer through a α-helix; protein located entirely in the cytosol and anchored to the cytosolic monolayer through covalent lipid chains; peripheral protein attached to the inner side of the membrane and peripheral protein attached to the outer side of the membrane); glycolipids and glycoproteins.

In order to make the modeling structures accurate and reliable, when creating them, we used as reference the images present in the books "Molecular Biology of the Cell" (ALBERTS et al., 2017) and "The Cell: a molecular approach" (COOPER; HAUSMAN, 2007), both present in the basic bibliography of Biological Sciences courses and health area courses.

The different types of lipids and proteins, glycoprotein and glycolipid were modelled separately, and then organized according to the fluid mosaic model of the plasma membrane structure, as shown in figure 8 (page 34) presented in the section on Teaching Cell Biology in Chapter I of this research. The process of modeling each cell structure will be described below.

a) Cholesterol

Cholesterol is a sterol that contains a rigid ring structure attached to a single polar hydroxyl group and a small apolar hydrocarbon chain (ALBERTS et al., 2017). In the modeling process, the spatial and chemical formula was considered, as well as its didactic scheme, as illustrated in figure 13.

Figure 13 - Cholesterol Structure

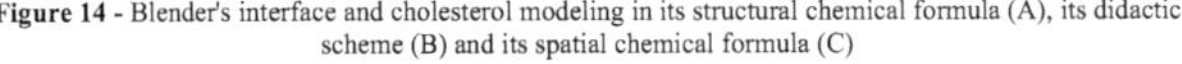

Chemically, cholesterol consists of 27 carbon atoms, 46 hydrogen atoms and 1 oxygen atom, which in the modeling were represented by spheres in the colors black, white and red, respectively. The modeling process (figure 14) was based on the model highlighted by Alberts et al. (2017), figure 13.

Figure 14 - Blender's interface and cholesterol modeling in its structural chemical formula (A), its didactic scheme (B) and its spatial chemical formula (C)

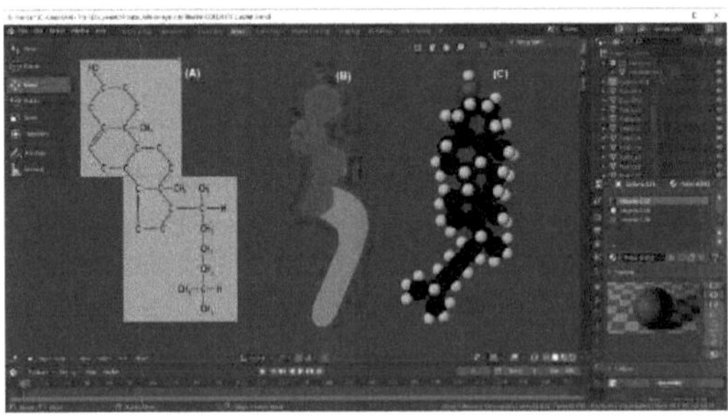

b) Phospholipids

Phospholipids are the most abundant lipids in the plasma membrane. They consist of a polar head group containing one phosphate group and two hydrophobic hydrocarbon tails, one of which has instabilities, i.e. double bonds between the carbons (COOPER; HAUSMAN, 2007), as shown in figure 15.

Figure 15 - Parts of a typical phospholipid molecule

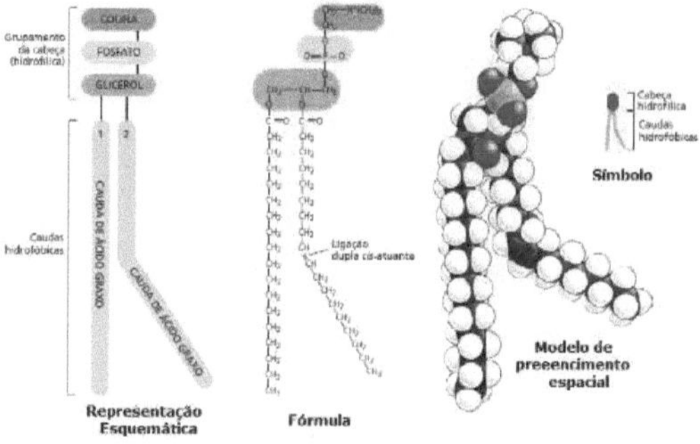

In animal cells, the plasma membrane consists mainly of phosphoglycerides, which have a main three-carbon ($C_3H_5O_3$) glycerol chain. The carbons adjacent to the glycerol bind in the two long hydrocarbon chains and the third carbon atom of the glycerol binds to a phosphate group. This in turn binds to various types of groups, thus constituting different phosphoglycerides and the most abundant in the animal cell plasma membrane are: phosphatidylethanolamine, phosphatidylserine, phosphatidylcholine and phosphatidylinositol (ALBERTS et al., 2017).

Another important class of phospholipids are sphingolipids, and instead of glycerol, they present sphingosine, which is a long acetyl chain with an amino group (NH_2) and two hydroxyl groups (OH) at one end. In these phospholipids, a hydrocarbon tail is bound to the amino group and a phosphocholine group is bound to the end hydroxyl group, as is the case of sphingomyelin (ALBERTS et al., 2017).

The phospholipids were modeled in three different ways: their symbol, their structural chemical formula and their spatial chemical formula. The symbol, as can be seen in figure 15, consists of a hydrophilic head and a hydrophobic tail, so that the phospholipids could be identified, the heads were represented by different colours. And the hydrocarbon tails were also presented in different colours to differentiate phospho glycerides from sphingolipid, as shown in Table 3.

Table 3 - Identification of Phospholipidic Didactic Representations

PHOSPHOLIPIDE	POLAR HEAD COLOUR	HYDROCARBONATE TAIL COLOUR	MEMBRANE LOCATION
Phosphatidylinositol	Dark Green	Rosa	Internal Monocamada
Phosphatidylethanolamine	Yellow	Rosa	Internal Monocamada
Phosphatidylserine	Light Green	Rosa	Internal Monocamada
Phosphatidylcholine	Red	Rosa	External Monoclayer (predominantly)
Sphingomyelin	Brown	Red	External Monoclayer (predominantly)

Source: author, 2018.

The modelling of the spatial and structural chemical formulas of phospholipids was developed from the models present in (COOPER; HAUSMAN, 2007), as shown in figure 16. In the modeling of the spatial formulas, spheres of different colors and sizes were used, representing the atoms that constitute them, and transformations were applied to the spheres, of scale, rotation and translation to achieve the desired shape. The atoms that constitute phospholipids are carbon, hydrogen, phosphorus, nitrogen and oxygen, which were represented by the colors black, white, green, blue and red, respectively.

Figure 16 - Structure of phospholipids (spatial and structural formulas)

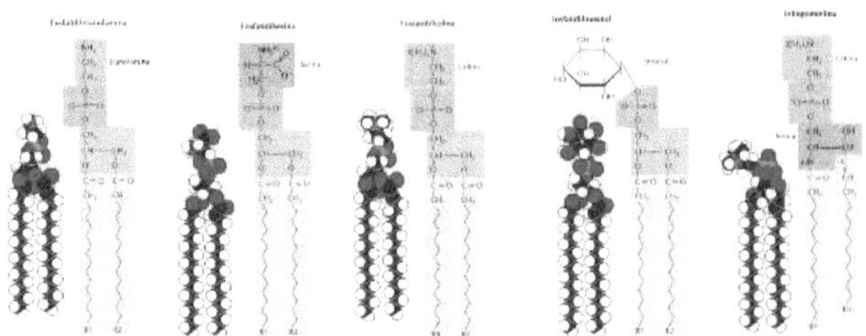

Source: COOPER; HAUSMAN, 2007.

According to ALBERTS et al. (2017), the angles that the atoms establish between themselves at the time of the chemical bonding, therefore there is a slope of approximately 45° in one of the hydrocarbon chains of phospholipids, because it represents the presence of an unsaturation, i.e., a situation in which the carbon performs a double bond, which causes a change in the angulation of the chemical bond established between these atoms, highlighted in Figure 15.

Figures 17, 18, 19, 20 and 21 represent the modelling of the symbol, the structural chemical formula and the spatial chemical formula of phospho glycerides (phosphatidylethanolamine, phosphatidylserine, phosphatidylcholine and phosphatidylinositol) and sphingolipid (sphingomyleline).

Figure 17 - Blender's interface and modeling of phosphatidylethanolamine in symbol form (A), its structural chemical formula (B) and its spatial chemical formula (C)

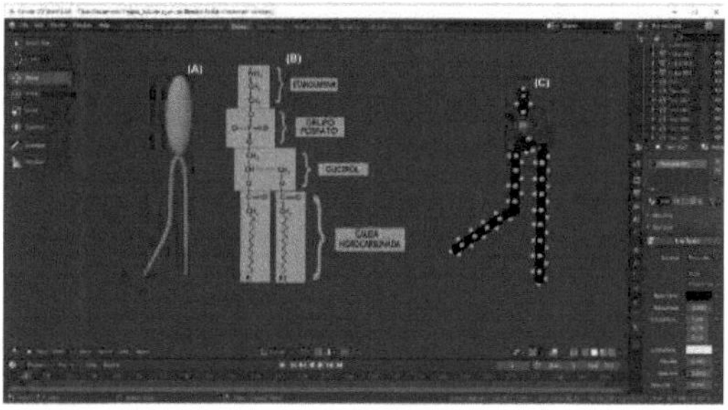

Source: author, 2019.

Figure 18 - Blender's interface and Phosphatidylserine modeling in symbol form (A), its structural chemical
formula (B) and its spatial chemical formula (C)

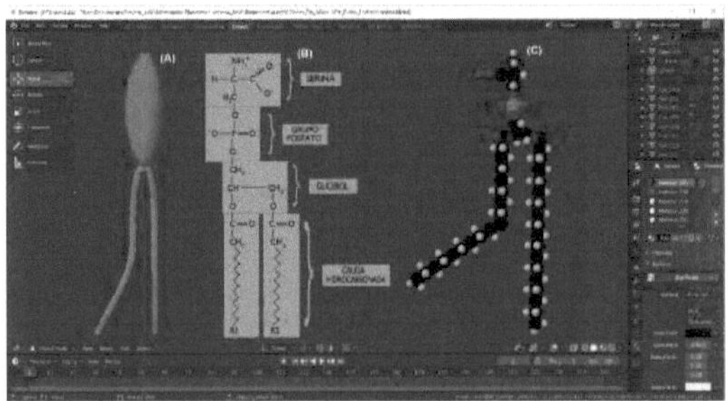

Source: author, 2019.

Figure 19 - Blender's interface and Phosphatidylcholine modeling in symbol form (A), its structural chemical
formula (B) and its spatial chemical formula (C)

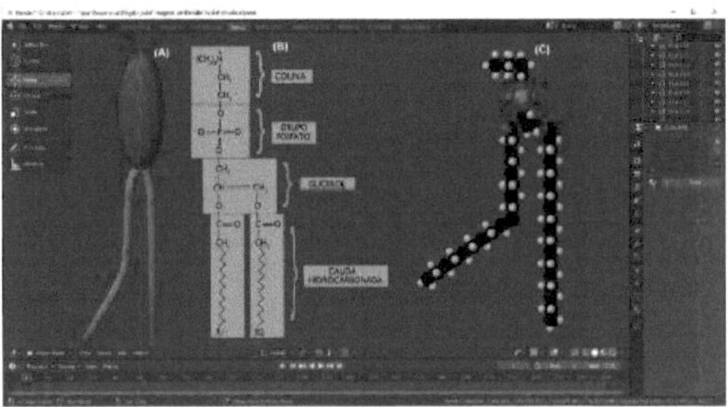

Source: author, 2019.

Figure 20 - Blender's interface and modeling of Phosphatidylinositol in symbol form (A), its structural chemical formula (B) and its spatial chemical formula (C)

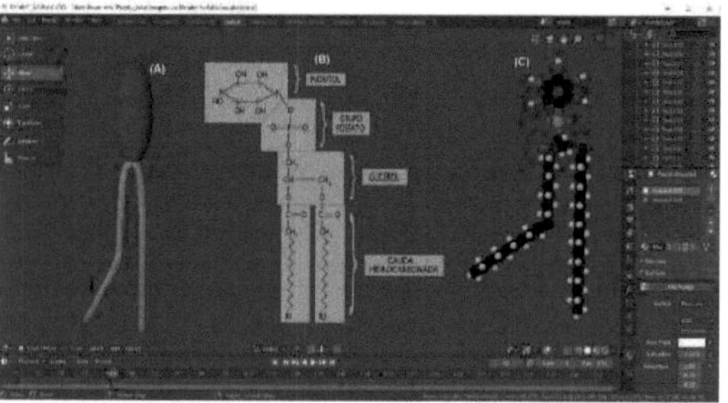

Source: author, 2019.

Figure 21 - Blender's interface and Sphingomyelin modeling in symbol form (A), its structural chemical formula (B) and its spatial chemical formula (C)

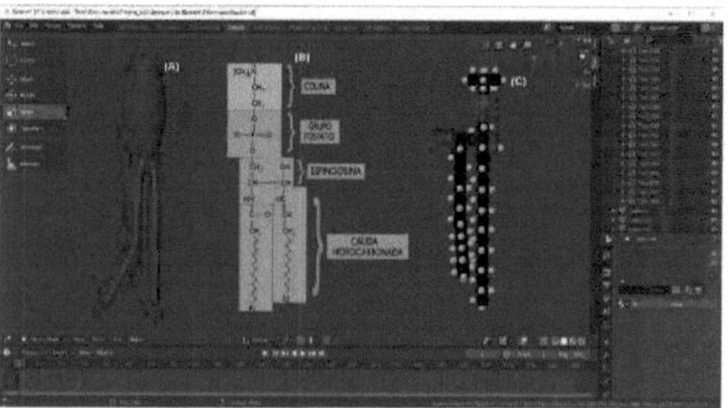

Source: author, 2019.

c) Membrane Proteins

Membrane proteins are two-dimensional, fluid and perform specific functions on the plasma membrane, and are inserted into the lipid layers. There are two classes of proteins associated to the membrane: peripheral membrane proteins, which can be dissociated from the plasma membrane through the use of polar reagents or with high concentrations of salts; and integral membrane proteins, which are inserted in the lipidic layer and are only dissociated through the use of reagents capable of breaking hydrophobic interactions (COOPER; HAUSMAN, 2007).

In order to model the proteins, it was necessary to understand a little more about *Bézier*'s Curves and *Bevel Object*, which allow to delimit the shape that the modeling should follow. The representation chosen for protein modeling was cylindrical, resembling the proteins represented by (ALBERTS et al., 2017), which served as the basis for building the plasma membrane model (figure 22).

Figure 22 - Several ways in which proteins are associated with the lipid layer

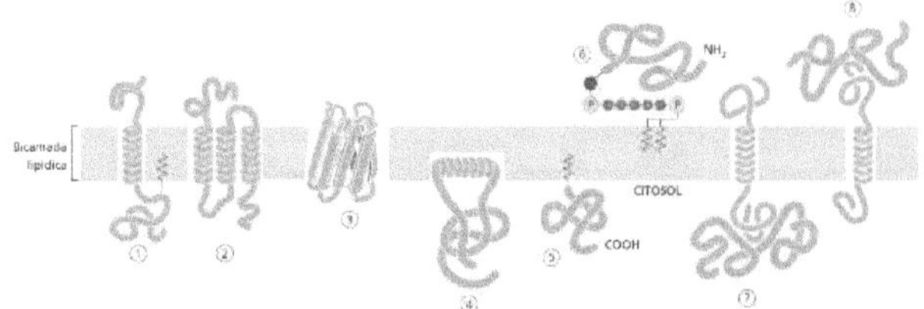

Source: ALBERTS et al., 2017.

Transmembrane proteins are those that cross the lipidic layer and have hydrophobic and hydrophilic regions. The hydrophobic region crosses the membrane and interacts with the hydrocarbon tails of phospholipids, keeping them out of the water. The hydrophobic regions are exposed to water on both sides of the membrane (ALBERTS et al., 2017). The following transmembrane proteins have been modelled: transmembrane protein with a single α-helix (figure 23A); transmembrane protein with multiple α-helix (figure 23B); transmembrane protein with a leaf β (figure 23C); protein located entirely in the cytosol and attached to the cytosolic monolayer through a α-helix (figure 23D); protein located entirely in the cytosol and attached to the cytosolic monolayer through covalent lipid chains (figure 23E); and protein totally exposed to the external surface of the cell (figure 23F).

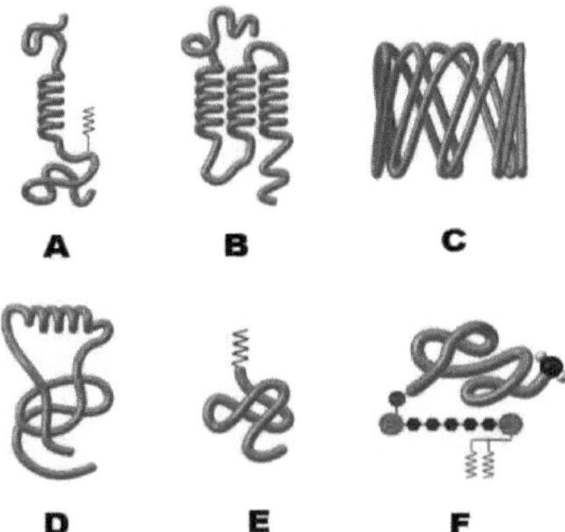

(A) Single-screw transmembrane α-helix; (B) Multi-screw transmembrane α-helix; (C) Single-screw transmembrane β; (D) Protein located entirely in the cytosol and attached to the cytosolic monolayer by a α-helix; (E) Protein located entirely in the cytosol and attached to the cytosolic monolayer by covalent lipid chains; (F) Protein totally exposed to the external surface of the cell.

Source: author, 2018.

In the constitution of the plasma membrane, there are also the proteins associated with the membrane, which are those that do not extend into the hydrophobic interior of the lipidic layer, its connection to the plasma membrane occurs on one of the faces, internal or external, of the membrane, through non-covalent interactions with other proteins of the membrane (ALBERTS et al., 2017). The peripheral protein bound to the internal side of the membrane and the peripheral protein bound to the external side of the membrane were modeled (figure 24), their differences will only be noticeable in the plasma membrane model (figure 29).

Figure 24 - Rendering of membrane associated proteins.

Source: author (2018).

d) Glycolipids

Like phospholipids, these compounds are formed by a hydrophobic region containing two long hydrocarbon tails and a polar region, but instead of a phosphate group attached to the head, it has one or more sugars (ALBERTS et al., 2017).

The modeling of glycolipids (figure 25) was similar to what was done in the modeling of the phospholipids symbol, what differentiated it was the head, which in glycolipids has hexagonal shape.

Figure 25 - Glycolipid Rendering

Source: author, 2018.

e) Glycoproteins

Glycoprotein is a type of protein that has a carbohydrate attached to it, the binding process of this carbohydrate occurs during the translation of the protein or as a post-translation modification in a process called glycosylation (ALBERTS et al., 2017).

The same technique was used for modelling this structure (figure 26) as for the other proteins. Then, hexagons were created to represent the glycides, which were joined to proteins, thus constituting the glycoproteins.

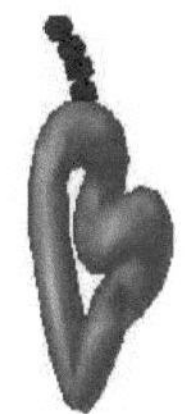

Figure 26 - Glycoprotein yield

Source: author, 2018.

f) Plasma Membrane

For the representation of the plasma membrane in the virtual environment, the asymmetric distribution of phospholipids and glycolipids in the lipidic layer was taken into consideration (figure 27), and two representations were chosen, one in fluid mosaic and the other "didactic", demonstrating only one layer of phospholipids and highlighting several types of transmembrane proteins, aiming to provide a better understanding and better interaction in the virtual environment.

Figure 27 - Asymmetric distribution of phospholipids and glycolipids on the lipid layer

Source: ALBERTS et al., 2017.

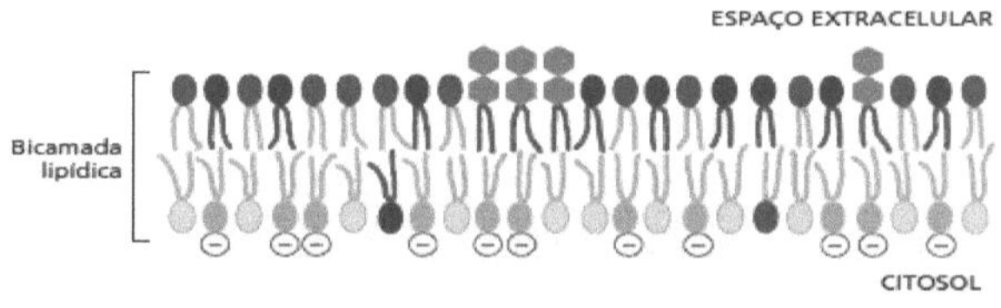

In the process of modeling the plasma membrane in fluid mosaic, the phospholipid layers were duplicated several times, modifying the position of the set created in the Z axis and slightly in the Y axis. Then, glycolipids, some transmembrane proteins and cholesterol were inserted, thus constituting the plasma membrane (figure 28).

Figure 28 - Blender interface and modeling of the fluid mosaic plasma membrane

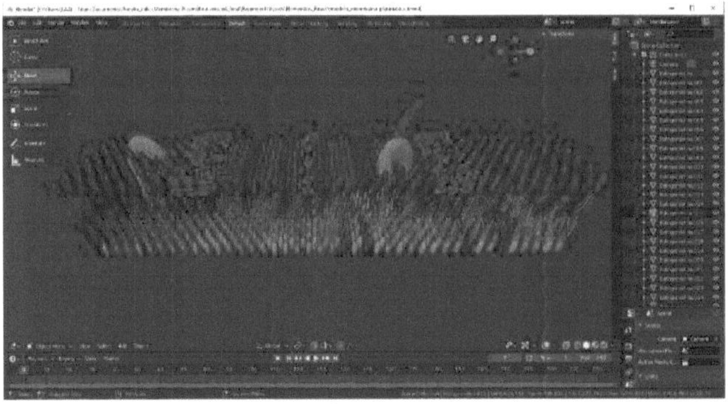

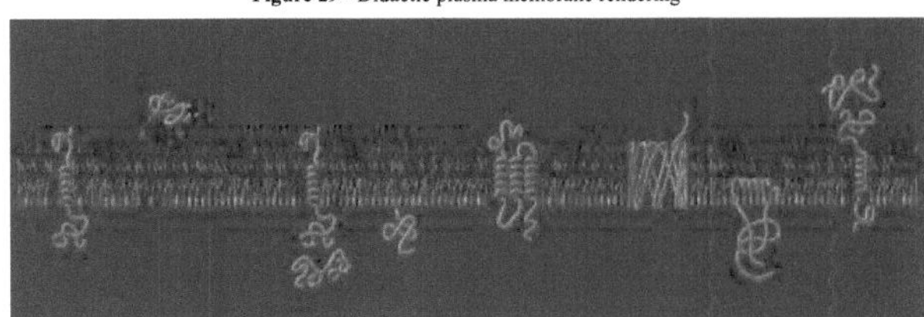

Source: author, 2018.

Figure 29 shows the "didactic" model of the plasma membrane. For its construction, the symbol of phospholipids and cholesterol was modelled; glycolipids and various types of proteins, all aligned on the X, Y and Z axes. The asymmetry of phospholipids was respected, as well as the colors of those that are turned to the external environment and those that are turned to the cytosol (table 3). Enough spaces were left on the membrane to insert the various types of proteins created, and the instances of cholesterol where the protein was inserted had to be removed, since this is what occurs in a real plasma membrane.

Figure 29 - Didactic plasma membrane rendering

Source: author, 2018.

g) Molecules that participate in Membrane Transport

The internal composition of the cell is maintained because the plasma membrane is selective and permeable to small molecules, i.e. most biological molecules are able to diffuse through the lipid layer, so that the plasma membrane forms a barrier to block the free exchange of molecules between the cytoplasm and the external environment of the cell (COOPER; HAUSMAN, 2007).

This passage of molecules through the plasma membrane determines the transports, which can be passive (simple diffusion, facilitated diffusion, ionic channels) or active (sodium and potassium pump). These transports are simulated in the developed virtual environment, so it was necessary to model some of these molecules.

The oxygen (figure 30A) and carbon gases (figure 30B) were modeled; the hydrophobic molecule, benzene (figure 30C); the small polar molecules, water (figure 30D) and ethanol (figure 30E); the large polar molecule, glucose (figure 30F) and the charged molecules: amino acids (figure 30G), sodium ion (Na+) (figure 30H) and potassium ion (K+) (figure 30I).

In the modeling of all these molecules, spheres of different colors and sizes were used, representing the atoms that constitute them. And in order to acquire the desired shape, transformations were applied to the spheres, of scale, rotation and translation. The atoms that constitute the modeling molecules are carbon, hydrogen, nitrogen, oxygen, sodium and potassium, which were represented by the colors black, white, blue, red, purple and orange, respectively.

Figure 30 - Rendering of biological molecules that participate in transport through the membrane

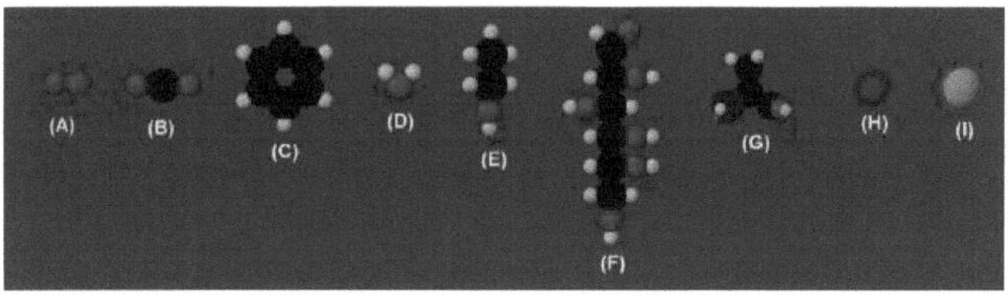

Source: author, 2019.

h) Carrier Proteins and Ionic Channels that Allow the Passage of Molecules

The transport of substances through the plasma membrane may occur by the phospholipidic layer or mediated by carrier proteins and ionic channels, also called protein channels. The carrier proteins bind selectively and carry small specific molecules, such as glucose. Therefore, they act as enzymes to facilitate the passage of specific molecules through the membrane. By binding to these molecules, they undergo conformative changes that open the channels through which the molecule to be carried can cross the membrane and be released on the other side (figure 31A). The ionic channels form open pores through the membrane, allowing the free diffusion of any molecule with appropriate size and load. The pores formed by these protein channels are not permanently open, but can be selectively opened or closed in response to extracellular signals, allowing the cell to control the movement of ions through the membrane, figure 31B (COOPER; HAUSMAN, 2007).

Figure 31 - Carrier Protein and Protein/Channel

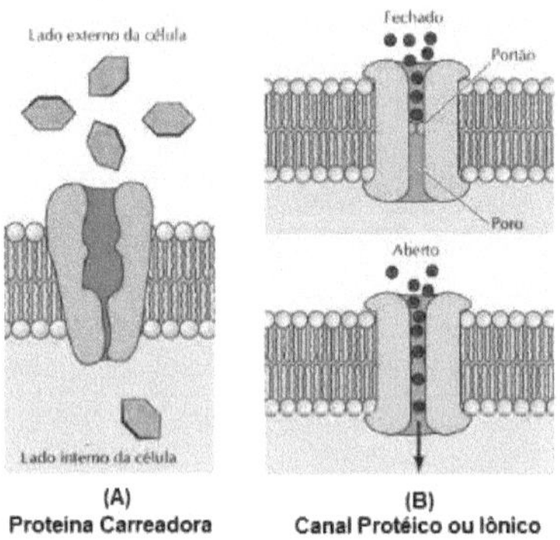

Source: COOPER; HAUSMAN, 2007.

To simulate the transports through the plasma membrane, it was necessary to model the carrier protein (figure 32) and the ionic channels (figure 33), which participate in the passive (figure 34) and active transports (figure 35), demonstrated in the simulator.

Because it is a complex object, the modeling of these structures followed the pattern of protein modeling, with the use of the *Nurbs Circle*, which allows the modification of the modeling to the format you want.

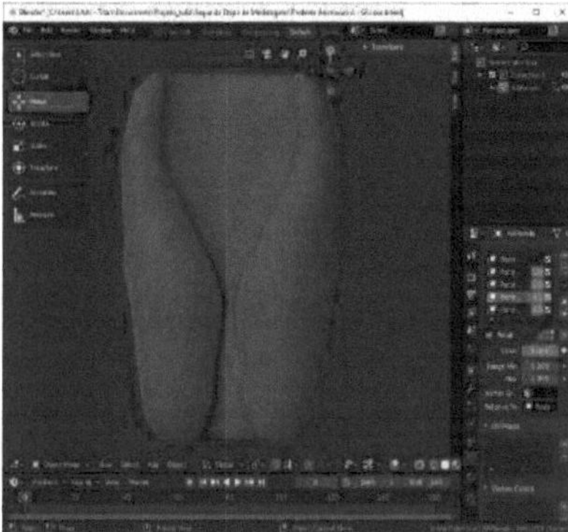

Source: author, 2019.

Figure 33 - Blender interface and ionic channel modeling

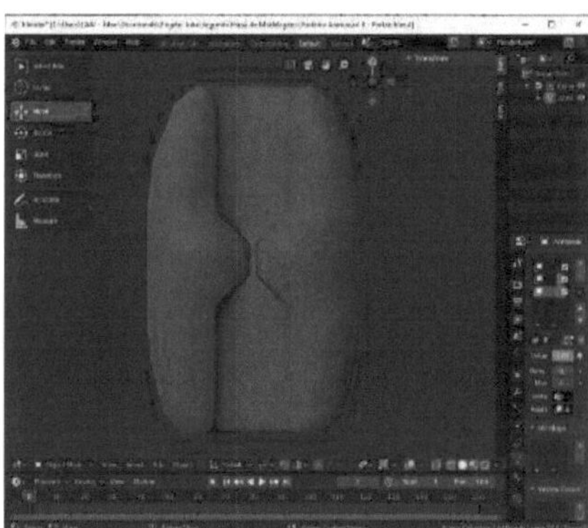

Source: author, 2019.

Figure 34 - Blender interface and potassium ion channel modeling

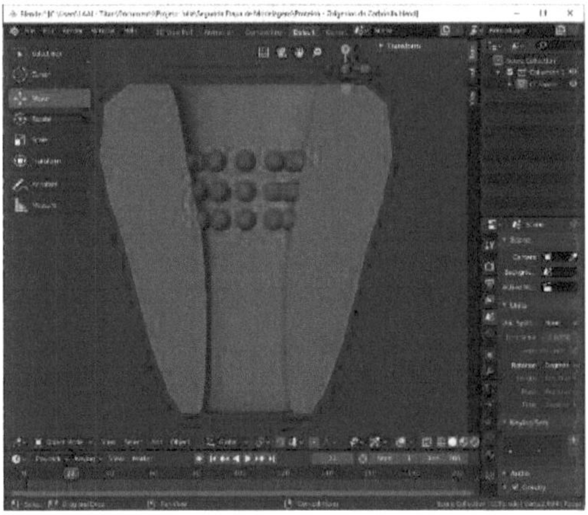

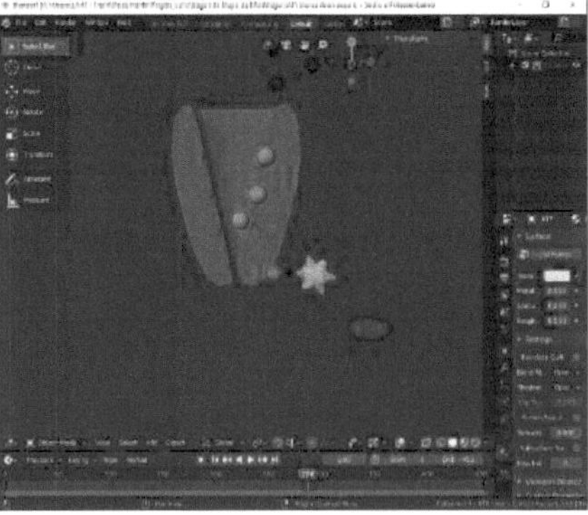

Source: author, 2019.

Figure 35 - Blender interface and modeling of the protein that participates in active transport

Source: author, 2019.

i) Structure of the Glucose Conveyor

Among the various types of carrier proteins existing in the human species, the glucose carrier was the best studied. Initially, this carrier was identified as a 55 kD protein in human red blood cells, where it represents approximately 5% of total membrane proteins. Later, through isolation and sequential analysis of cDNA clones[12], they concluded that the glucose carrier is a α-helix transmembrane protein that exceeds the bilipidic layer twelve times (figure 36).

Figure 36 - Structure of the Glucose Carrier

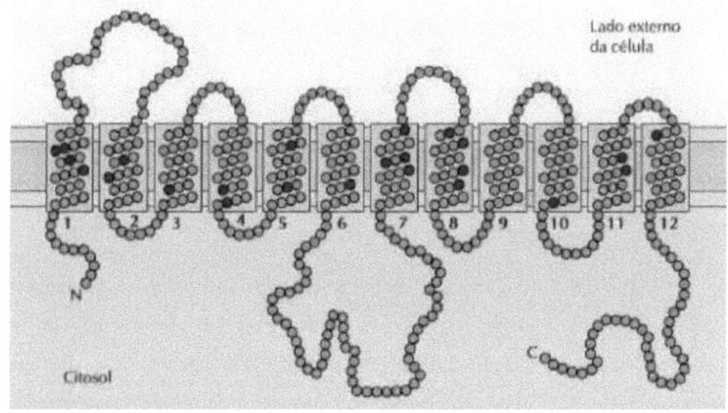

Source: COOPER; HAUSMAN, 2007.

Because it is an important carrier protein for the human species, it was thought necessary to model it so that the student could understand that the carrier protein and ion channel models present in the simulator represent a didactic model for understanding its process, and does not necessarily show its chemical structure.

In the process of modelling the glucose carrier (figure 37), the *Bezier* curve was used, which allows editing in the desired format.

---

[12] It is the DNA synthesized from a messenger RNA molecule, whose introns have already been removed, that is, the mRNA has already gone through the process of *splicing*, being a reaction catalyzed by the reverse transcriptase enzyme.

Figure 37 - Blender Interface and Glucose Conveyor Modeling

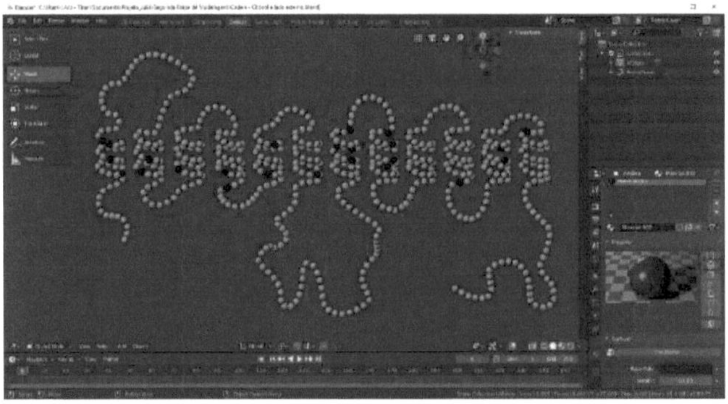

Source: author, 2019.

j) Animal Cell

After the development of the structures that will compose the simulator in a VR environment of the plasma membrane, it was thought necessary that the user would need to have an overview of the cell and not only the plasma membrane. Then, the modeling of an animal cell was developed (figure 38), with its main cytoplasmic organelles (figure 39), in which the user can interact with all the cellular structures and perform the simulation of the transport of substances through the plasma membrane.

Figure 38 - Blender Interface: External cell view

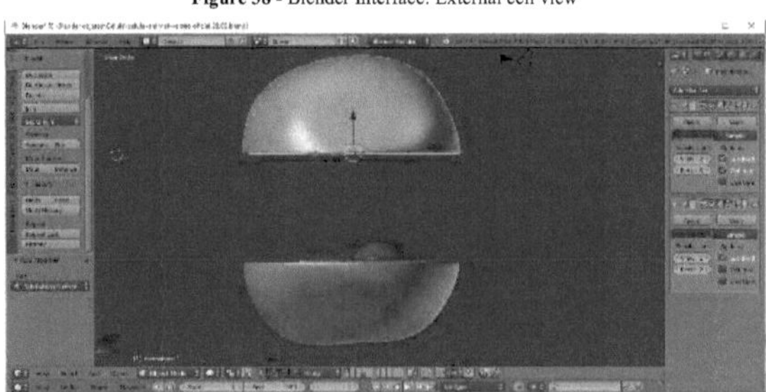

Source: author, 2019.

Figure 39 - Blender's interface: Internal view of the cell showing the cytoplasmic organelles

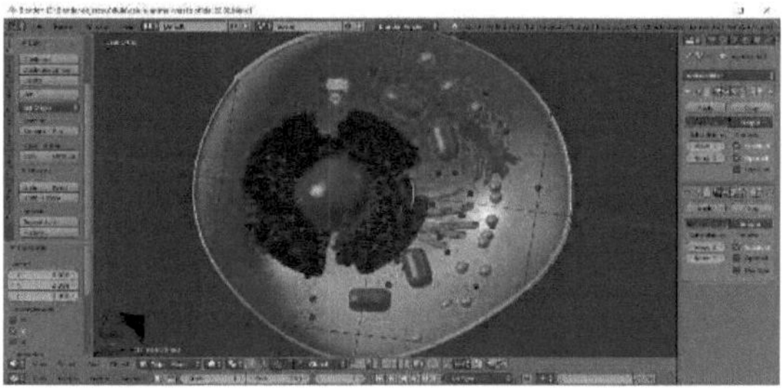

Source: author, 2019.

Once the modeling of cell structures was concluded, it was clarified for the team how the movements of phospholipids and the transport of substances through the plasma membrane to be developed in the simulator occur, and it was directed to the integration stage to the VR environment.

4.2 Integration into the VR Environment

The essence of VR is in the creation of simulated worlds, totally built by the computer, with which the participant can interact directly and in real time, discovering and learning through sensations coming essentially from sight, hearing and touch.

We all learn, above all, by doing; it is through practice, through first-hand experience, that we discover reality and learn to know it. In this sense, VR has multiple attributes and specificities, which, if used correctly in the area of education, represents an excellent tool for teaching learning, bringing together all the elements that an efficient didactic requires: teaching through the senses; integrating theory with practice; transforming the complex into the simple; transforming the abstract into the real. This allows learning by one's own experience, and can become a powerful teaching tool for teachers in all areas (CAMACHO, 1996).

The use of simulators in VR is recommended for many formalized and abstract contents, because the visual representations favor understanding and learning. The use of VR simulators in education represents an excellent tool to facilitate learning, because it allows the student to have a deep interaction with the content, manipulating and obtaining, with this activity, an active and enriching experience that will allow them to efficiently and lastingly understand the content (EICHLER; PINO, 2006).

After modeling all the structures needed to build the virtual environment, these were imported into the *Unity* 3D game development engine. Since the *HTC VIVE Oculus* RV goggles (figure 40) used in the simulator do not have a specific program to handle them and create RV environments, it was decided to use Unity *3D* to integrate *VIVE* into the developed environment. The immersion of the student in the virtual environment of the plasma membrane is carried out by using the *HTC VIVE Oculus hardware*, consisting of the glasses, controllers and base station. A complete virtual reality equipment, featuring a complex lens and technology system; detectors with internal and external circulation, allowing the user a real virtual experience without limits. With 32 sensors installed, it allows a vision, without blind spots, and the 360 cover allows the locomotion with total freedom. The glasses have adjustable straps (figure 40B); interchangeable inserts; front camera (figure 40A); and eye relief adjustments that provide comfort to the user, because the glasses fit perfectly to your face (Vive s.d., VIVE ™ | Virtual Reality System 2018).

The base station is responsible for immersing the RV, which is done by capturing the signal emitted by a source, which is then processed by a control box and sent to the computer, and, from an initial reference point, measures changes in orientation and position of the mechanism where the sensor is installed, this allows tracking coverage of the 360-degree playback area through wireless synchronization. The configuration of these stations is quite simple, you only need one power cable to run. Each controller has twenty-four sensors that enable users' actions to be wirelessly and faithfully recreated in the virtual world where it is inserted, allowing them to interact (VIVE, 2018).

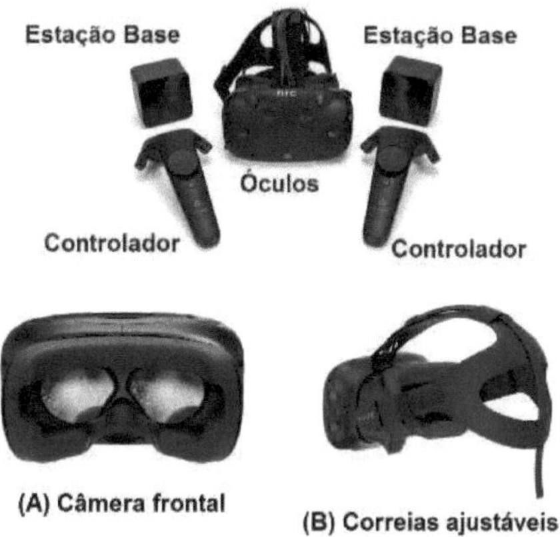

Installing the *HTC VIVE Oculus* requires space, so that the RV goggles work in the best possible way. An open area of at least 1.5 metres wide by 2 metres high is recommended. The sensors that will recognize the device in the environment must not be more than 5 meters away from each other, 2 meters high and must be angled 30 to 45 degrees while perpendicular to the wall, otherwise the device will not operate smoothly (figure 41A). In addition, the computer that will work in conjunction with the *headset will* also be in a field of view of up to 5 metres. Base stations do not need to be connected to the *HTC VIVE* or your computer to actively exchange data. The *headset* identifies invisible light pulses emitted by the two towers to determine its position on site, so that the *HTC VIVE* recognizes the user and his movements, the entire environment needs to be clean, i.e. without any obstacles around, figure 41B (VIVE, 2018).

Figure 41 - *HTC VIVE Oculus* installation

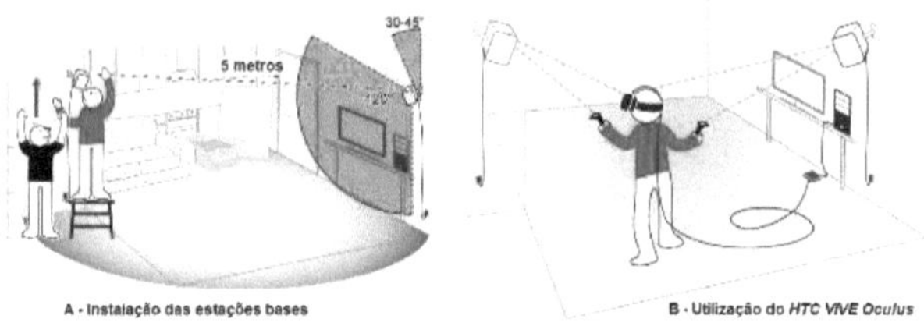

Source: VIVE, 2018.

The development of the simulator was initiated after a process of awareness and explanation by the author of the work to the production team on the dynamics of phospholipids movements in the plasma membrane, and on how small molecules transport through the plasma membrane. The first step was to define the commands to be determined in the control, for the movement and user interaction in the cellular environment.

It was defined that the button called "trigger" would be used for the user to move in the virtual environment, the *touchpad* button, responsible for selecting and manipulating the structures, and the *Grip button* would allow the change of visualization of phospholipids and cholesterol within the Plasma Membrane Museum; and the repetition of processes in environments that simulate the transport of substances through the plasma membrane (figure 42).

Figure 42 - *HTC VIVE Oculus* controls

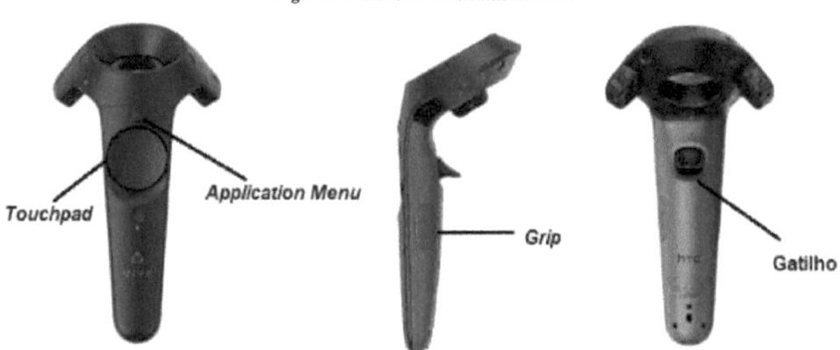

Source: VIVE, 2018.

70

It was defined that, during navigation, the user would have total freedom to move around the cellular environment and explore its structures. Then, the position of the plasmatic membrane and the molecules that participate in the transport of substances through the membrane were established and their interactions were programmed. It was also established commands that allow the user to participate in the transport of substances through the plasma membrane.

It was thought necessary to build a tutorial (figure 43), with a legend of the atoms and ions that make up the structures present in the simulator, so that the user can become familiar with the virtual environment.

Figure 43 - Visualization of the Tutorial in the VR environment

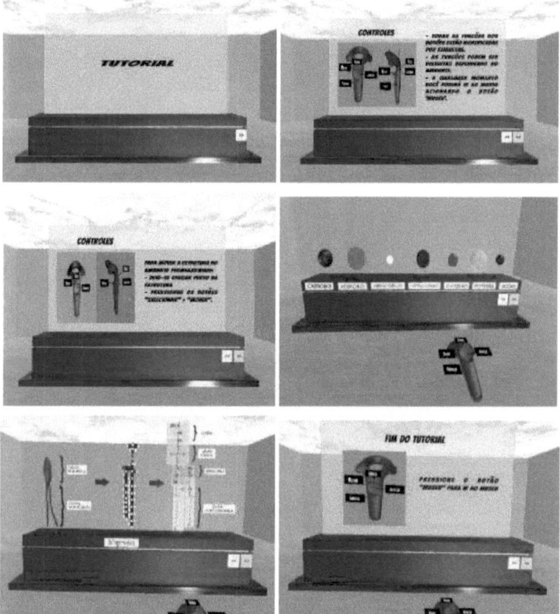

Source: author, 2019.

In addition, the tutorial explains to the user how the controller buttons work, after all, they are essential for the user to feel immersed in the RV environment and interact with the structures present in the environment. Therefore, it is important that the user understands their commands and learn how to handle them so that possible difficulties during navigation are minimized. Some environments have differentiated interaction commands, so, to guide the user, ID cards indicating the functionality of each button according to the virtual environment in which it is inserted were added to the controls (figure 44).

Figure 44 - Visualization of the commands in the controllers of each RV environment

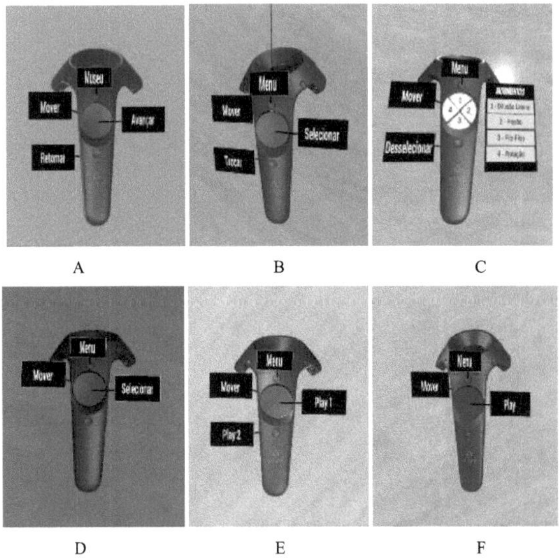

(A) tutorial control; (B) museum control; (C) phospholipids movement control; (D) simple diffusion control; (E) scene control: facilitated diffusion, ionic channel, Na+ channel, K+ channel, and active transport; (F) transport control of small molecules.

Source: author, 2019.

When entering the virtual environment, the user knows little about this world, but in a short time he realizes that this environment responds to his actions, presents new visual perspectives when he moves, responding in real time to his actions. For this student's experience in the virtual environment to be effective, it is interesting that they have a natural and intuitive behavior, facilitating navigation and interaction with the environment.

a) Plasma Membrane Museum

As soon as the guidelines of the tutorial are finished, the user is immersed in the Plasma Membrane Museum (figure 45), which aims to highlight the structures that make up the cell membrane, allowing the user to observe the three-dimensional molecular structures. Immersed in this environment, the user will be able to visualize the protein α single helix; the protein α multiple helix; the protein leaf β; the glucose carrier; the glycolipid; the cholesterol and phospholipids, all presented separately, on pedestals.

Cholesterol (figure 46) and phospholipids (figure 47) are presented in three different aspects, the symbol, the structural chemical formula and the spatial chemical formula, models that are commonly presented in Cell Biology textbooks. When entering the Museum, the user will visualize the symbol of the referred molecular compounds, to be able to visualize the other models it is necessary to press the control *grip* button (figure 42).

Figure 45 - Visualization of the Plasma Membrane Museum in the VR environment

Source: author, 2019.

Figure 46 - Visualization of Cholesterol in the Plasma Membrane Museum in the VR environment

Source: author, 2019.

Figure 47 - Visualization of Phospholipid (phosphotidilserine) in the Plasma Membrane Museum in the VR environment

Source: author, 2019.

The Plasma Membrane Museum is the first virtual environment of the simulator, besides visualizing the structures, the user has access to a *menu* (figure 48) that allows his passage to several other virtual worlds, which represent different learning environments. This passage occurs when the user selects, through the *touchpad* button of the control (figure 42), the environment to explore. As can be seen in figure 48, the simulator is composed of eight virtual environments. By selecting any of them, the user will be immersed in the cellular environment, establishing an interactivity relationship with the plasmatic membrane, and in the simulation has the capacity to act on the transport of substances through the plasmatic membrane and visualize them from different perspectives.

The following will detail what happens in each of these environments.

Figure 48 - Simulator main *menu* in VR environment

Source: author, 2019.

b) Phospholipid Movement Simulator

The plasma membrane is not a rigid and static structure, on the contrary, phospholipids are in constant lateral movement (lateral diffusion movement), providing a fluidity to the membrane. They can also rotate around their own axis (rotation movement) and present bending movements because of hydrocarbon chains. And very rarely performs the *flip-flop movement*, which is when it migrates from one single layer to another, this movement is rare because it requires that the polar portion of the molecule no longer interact favorably with water and cross the hydrophobic center of the bi-layer, a very endergonic process (NELSON; COX, 2019). All the movements mentioned are represented in figure 49.

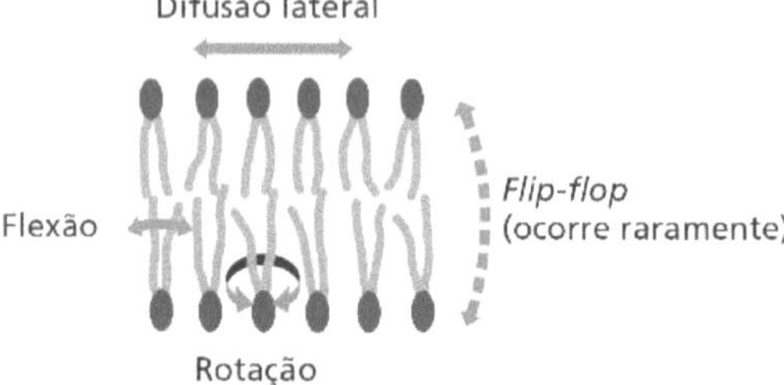

Figure 49 - The different movements of phospholipids in the plasma membrane

Source: ALBERTS et al., 2017.

In the first version of the simulator, these movements became difficult to be identified by the users because, as they occur simultaneously in the virtual environment, the user had difficulty differentiating each one of them (figure 50). Then, the way of visualization was modified, through commands in the control (figure 44C), which allows the user to select the movement he wants to visualize. After that, the phospholipids that are performing a different movement from the one selected acquire a neutral colour, thus highlighting the phospholipids that are performing the selected movement (figure 51).

Figure 50 - Simulator phospholipid movements in VR environment

Source: author, 2019.

Source: author, 2019.

c) Transport of Small Molecules through the Plasma Membrane

Due to its hydrophobic interior, the lipidic layer of cell membranes serves as a barrier to the passage of most polar molecules. This allows the cell to maintain different concentrations of solutes in the cytosol from those in the extracellular liquid and in each of the intracellular membrane-delimited compartments. However, to make use of this barrier, cells have developed means to transfer specific water-soluble molecules and ions through their membranes, after all, the cell is a living unit that needs to ingest essential nutrients, excrete toxic metabolic products and regulate intracellular concentrations of ions (ALBERTS et al., 2017).

One of the functions of the plasma membrane is to select the molecules that need to enter and/or leave the cell. This transport through the membrane is not simple and in some situations it is necessary to spend energy during the process. They are then classified as passive transport, when it does not involve energy expenditure and the movement of the molecules occurs from the most concentrated to the least concentrated medium; and in the active transport the movement of the molecules is against a concentration gradient, that is, it moves from the least concentrated to the most concentrated medium, and in order for this displacement to be possible, there is energy expenditure.

Virtual environments were then developed simulating the traffic of these small molecules both through the phospholipidic layer and by the aid of the transmembrane proteins that mediate this transport, which will be detailed below:

SIMULATOR OF SIMPLE DIFFUSION

Simple diffusion or passive diffusion occurs when small and relatively hydrophobic molecules (oxygen gas, carbon dioxide, benzene, water, ethanol) pass through the plasma membrane, through the phospholipidic bi-layer, dissolving in the aqueous solution on the other side of the membrane. Large unloaded polar molecules, such as glucose and charged molecules (ions), are unable to pass through the plasma membrane by simple diffusion (figure 52). In simple diffusion there is no participation of any protein, just as no external source of energy is involved and the direction of transport is determined by the relative concentration of the molecule inside and outside the cell, since its flow is always in the direction of the concentration gradient, being from the compartment of higher concentration to the one of lower concentration of the molecule, tending to promote the balance between the inside and outside side of the cell (COOPER; HAUSMAN, 2007).

Figure 52 - Permeability of the phospholipidic layer

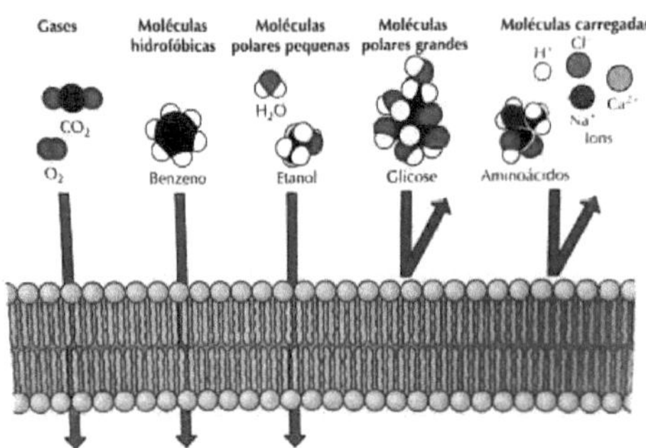

Source: COOPER; HAUSMAN, 2007.

In the simple diffusion simulator, the user will have a view of the external environment with the presence of a cell and, close to it, the molecules of water, ethanol, benzene, glucose, amino acid, oxygen and carbon dioxide are found.

The user will need to perform the simple diffusion and, to do so, needs to choose one of the molecules present in the middle, get close enough to it so that it is surrounded by a yellow coloration and then select the molecule by pressing the *touchpad* button of the controller (figure 42). Thus, you can move towards the cell by pressing the trigger of the controller. If you have selected a molecule (water, ethanol, benzene, oxygen gas and carbon dioxide) capable of dissolving in the phospholipidic layer, it will enter the cell together with the user. On the other hand, if the user selects a large molecule (glucose, amino acid) incapable of overtaking the cell through the phospholipidic layer, the cell will be involved by a red coloration and will distance itself from the cell (figure 53).

Figure 53 - Highlight of Simple Diffusion in the simulator in VR environment

Source: author, 2019.

SIMULATOR OF EASY DISSEMINATION

The facilitated diffusion, as well as the passive diffusion, occurs by the involvement of molecules in a direction determined by the relative concentrations on the inside and outside of the cell, with no participation of energy source.

What differentiates it from simple diffusion is the fact that the transported molecules do not dissolve in the phospholipidic layer. Its passage is mediated by proteins that allow the transport of molecules through the membrane without direct interaction with its hydrophobic interior (figure 54). Therefore, in this type of transport it is possible the passage of polar molecules and charged through the plasma membrane, such as carbohydrates, amino acids, nucleosides and ions.

The proteins that participate in the facilitated diffusion are the carrier proteins and the ionic channels. The carrier proteins are responsible for the passage of sugars, amino acids and nucleosides through the plasma membrane. (COOPER; HAUSMAN, 2007).

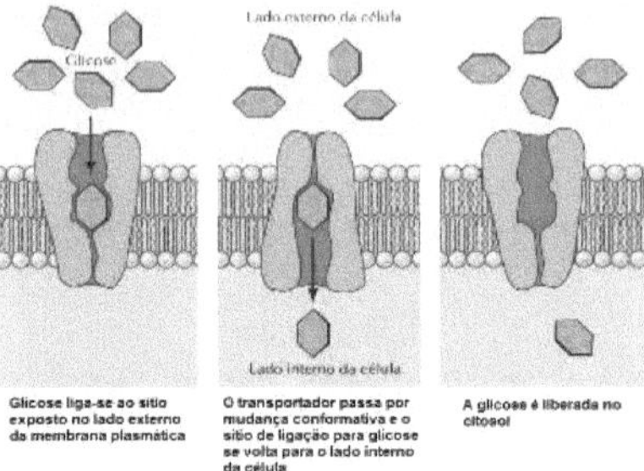

Figure 54 - Model for the facilitated diffusion of glucose

Source: COOPER; HAUSMAN, 2007.

A carrier protein carrying glucose was represented in the facilitated diffusion simulator (figure 55). When the user is immersed in the environment, to visualize the transport, he will need to press the *touchpad* button of the controller (figure 42), thus, he will be able to visualize the passage of glucose through the carrier protein, the same process will be repeated when pressing again the *touchpad* button of the controller (figure 42). If you want to see the passage of glucose through the carrier protein located on the other side of the environment, just press the *grip button of the* controller (figure 42), the process will be repeated by pressing again the *grip* button of the controller (figure 42).

Figure 55 - Featured Simple Diffusion Facilitated in the simulator in an RV environment

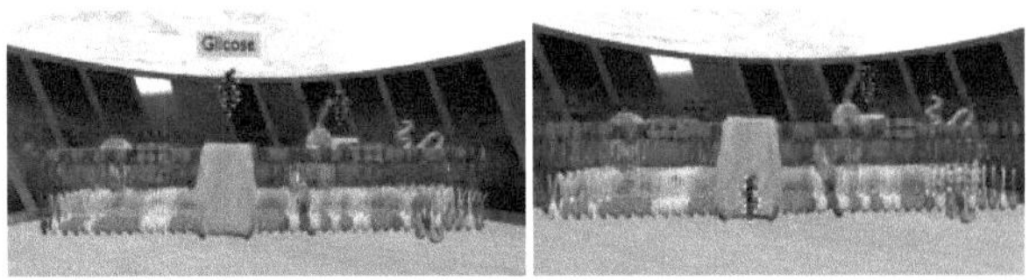

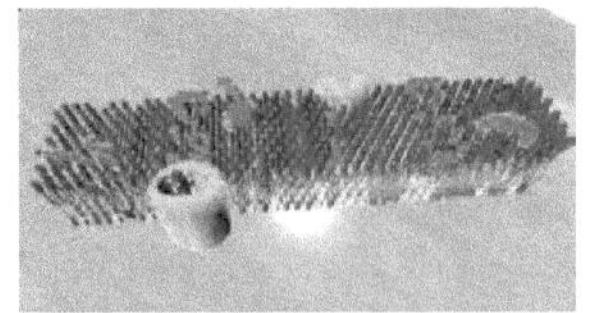

Source: author, 2019.

ION CHANNEL SIMULATOR

Ionic channels are present in the membranes of all cells and intermediate the passage of ions through the plasma membrane. This transport is extremely fast, with more than one million ions per second passing through the open ionic channels, which corresponds to a flow rate approximately one thousand times higher than the carrying rate by carrier proteins. These channels are highly selective, their narrow pores restrict the passage of ions with appropriate size and load, as well as allowing the passage of Na+, K+, Ca2+ and Cℓ- through the membrane. These channels are not permanently open, their opening and closing are regulated by specific stimuli, as can be seen in figure 31B, page 73 of this work (COOPER; HAUSMAN, 2007).

In the ion channel simulator (figure 56), the passage of the sodium ion (Na+) through the ion channel was represented. For the transport to occur, once immersed in the environment, the user can press the *touchpad* button of the controller (figure 42) to view the passage of ions through the channel, the same process will be repeated by pressing again the *touchpad* button of the controller (figure 42). If you want to view the transport through the ion channel located on the other side of the environment, just press the *grip button of the* controller (figure 42), the process will be repeated by pressing again the *grip* button of the controller (figure 42).

Figure 56 - Highlight of the ion channel in the simulator in RV environment

Source: author, 2019.

The highest ion selectivity index is given to the Na+ and K+ channels, therefore, it was thought important to simulate these channels so that the user can differentiate the process.

The selectivity of the sodium channel is due to the fact that the channel pore is very narrow, functioning as a filter by size (figure 57). The radius of the sodium ion is 0.95Å smaller than that of the potassium ion 1.33Å. The Na+ channel pore is narrow enough to prevent the passage of K+ or any other larger ion (COOPER; HAUSMAN, 2007).

Figure 57 - Na+ channel ion selectivity

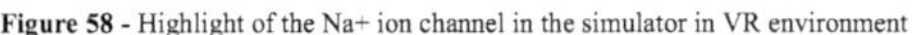

Source: COOPER; HAUSMAN, 2007.

In the sodium ion channel simulator (figure 58), the passage of the sodium ion (Na+) through the ion channel and the attempt to pass the K+ ion, which is repelled by the channel because it is too large, were represented. For the transport to occur, once immersed in the environment, the user will press the *touchpad* button of the controller (figure 42), thus, the passage of the sodium ion through the channel and the potassium ion being repelled, in case the process is desired to be repeated, just press the *grip* button of the controller (figure 42).

Figure 58 - Highlight of the Na+ ion channel in the simulator in VR environment

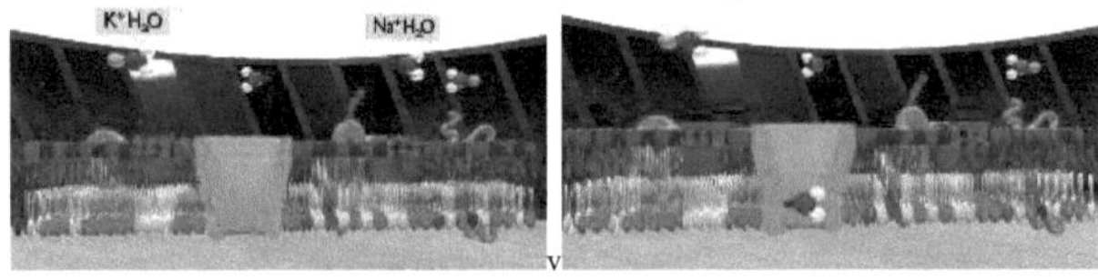

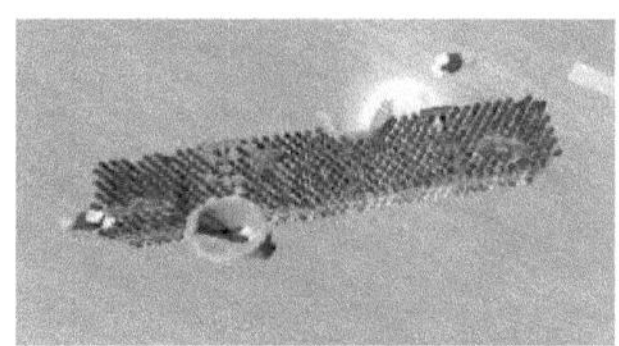

> ➤ **POTASSIUM ION CHANNEL SIMULATOR (K+)**

The potassium channels also have narrow pores, which prevent the passage of larger ions, but both Na+ and other smaller rays do not suffer the action of selective permeability by these K+ channels. This is because this channel has a selective filter, since it is internally surrounded by oxygen atoms of the carbonyl (C = O) of the polypeptide chain. The K+ ion, when in contact with this selective filter, interacts with the carbonyl oxygen and displaces the water molecule that was bound to the K+, thus the dehydrated potassium ion can overcome the pore. However, Na+ is very small and cannot interact with the carbonyl oxygen of the selective filter, so it remains bound to the water molecule and cannot overcome the channel, as can be seen in figure 59 (COOPER; HAUSMAN, 2007).

Figure 59 - K+ channel ion selectivity

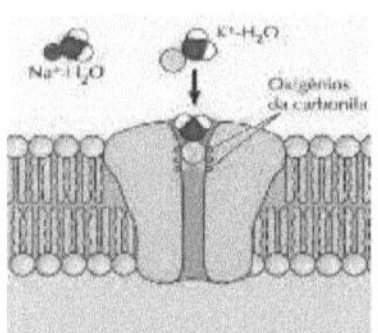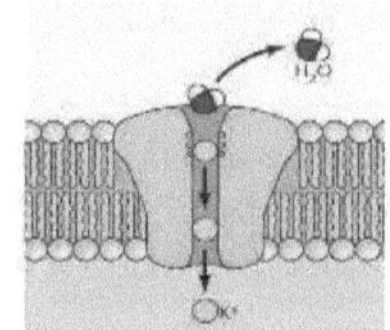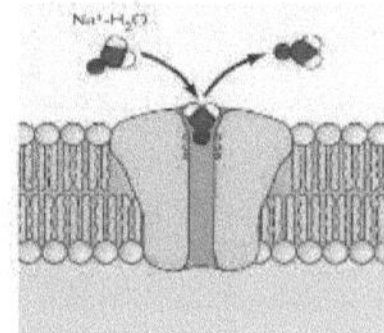

In the K+ ion channel simulator (figure 60), it is possible to observe the potassium ion attached to the water, approaching the ion channel. It then releases the water molecule, due to its interaction with carbonyl oxygen, and passes through the pore, reaching the interior of the cell. The sodium ion, even when combined with a water molecule, when approaching the ion channel, is repelled. For the transport to occur, once immersed in the environment, the user will press the *touchpad* button of the controller (figure 42), thus, will be able to visualize the passage of the ions through the channel, in case you want the process to be repeated, just press the *grip* button of the controller (figure 42).

Figure 60 - Highlight of the K+ ion channel in the simulator in RV environment

Source: author, 2019.

ACTIVE TRANSPORT SIMULATOR

Active transport occurs when the cell needs to transport a molecule against the concentration gradient, that is, it will be transported from where it is in smaller quantity to where it is in larger quantity. For this type of transport to happen, energy (ATP) is required. The sodium and potassium pump is an example of active transport. The concentration of Na+ is much higher outside the cell, while the concentration of K+ is higher inside. And these unequal ion gradients need to be maintained for proper functioning of nerve and muscle cells (COOPER; HAUSMAN, 2007).

The transport of sodium out of the cell and of potassium into the cell occurs as follows: three sodium ions that are inside the cell join the exposed site on the inner side of the cell. This bond stimulates ATP hydrolysis and pump phosphorylation, which induce conformative changes by exposing the Na+ binding site to the outer side of the cell and reducing its affinity for Na+, which is then released outside the cell.

And the binding sites of high affinity with K+ are exposed on the cell surface and two potassium ions bind to it, stimulating the hydrolysis of the phosphate group that is bound to the pump and inducing a second conformative change, which exposes the binding site from K+ to cytosol. This causes the binding affinity to reduce and the K+ to be released within the cell, as shown in figure 61 (COOPER; HAUSMAN, 2007).

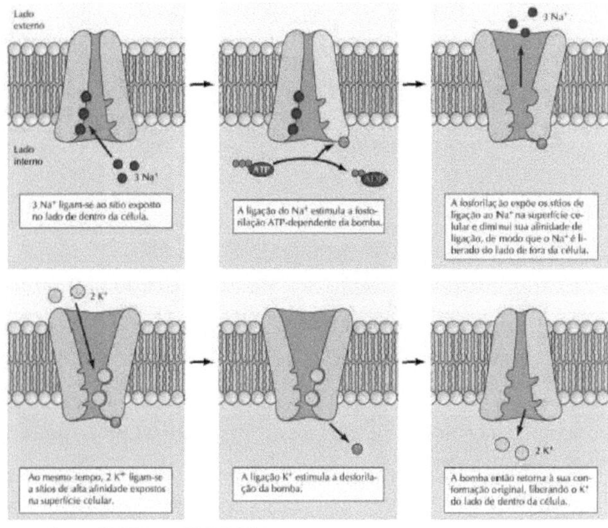

Source: COOPER; HAUSMAN, 2007.

In the simulator, the whole process was developed (figure 62), as soon as the user enters the virtual environment and press the *touchpad* button of the controller (figure 42), he will be able to visualize all the steps of the sodium and potassium pump, if he wants the process to be repeated, just press the *grip* button of the controller (figure 42).

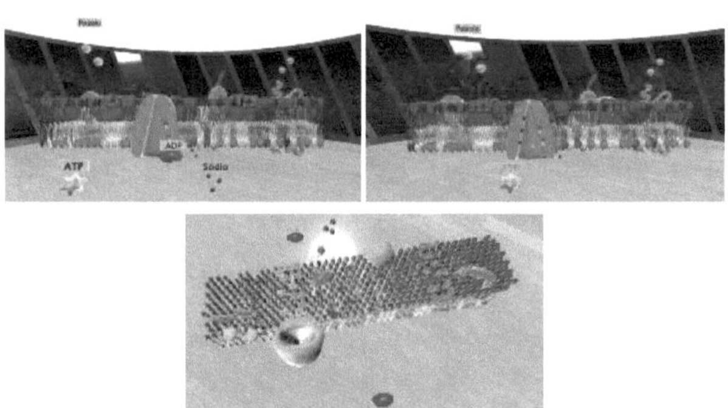
Figure 62 - Na+ and K+ pump highlight in RV environment simulator

Source: author, 2019.

SIMULATOR FOR THE TRANSPORT OF SMALL MOLECULES THROUGH THE PLASMA MEMBRANE

In this virtual environment, user involvement will be differentiated, you do not need to activate any action on the controller for it to occur. You will be able to visualize, simultaneously, all the processes described in the previous environments, having free navigation access, which allows an observation in different angles (figure 63).

Figure 63 - Highlight of the transport of small molecules in the simulator in VR environment

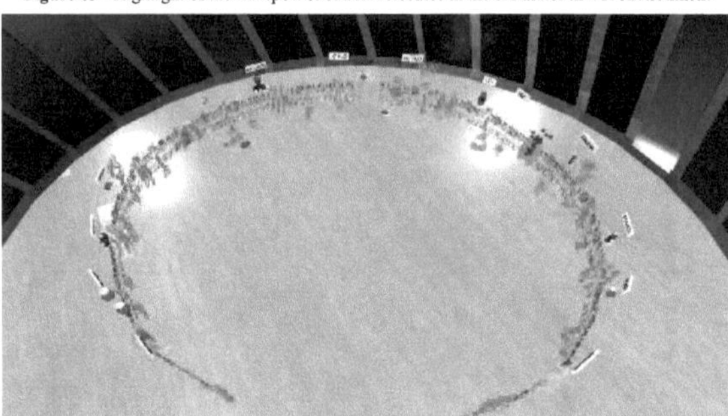

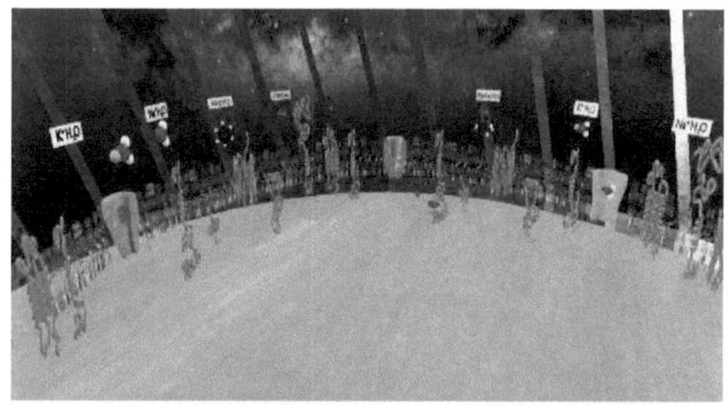

Source: author, 2019.

The virtual environments cited allow users to have first person experiences, and this is possible because the immersive VR eliminates the interfaces that act in the user-computer interaction. This allows a synthetic experience, which enables the user to obtain meanings close to what is obtained in the real world.

A very interesting factor of VR in education and teaching is the immersion in the virtual world, because it is difficult for students to have the opportunity to carry out their own experiences to draw their conclusions from the results obtained, which are then transformed into knowledge about the observed reality. The most common is the passive transmission of knowledge to the student, which makes their analysis and criticism difficult, since they do not have the necessary elements for that. But with the creation of virtual worlds, this personal experience becomes possible, because the student, entering the simulation, discovers for himself and learns by building his knowledge through the experienced VR. In this way, learning becomes richer, livelier, more varied and lasting, because, as it was constituted by a personal experience, it will hardly be forgotten (CAMACHO, 1996).

With the construction of this simulator in a virtual environment for the teaching of the plasmatic membrane, it was possible to make an abstract content, perceptible to the senses. This is because with the application, the plasmatic membrane gained form and consistency, being able to be visualized and manipulated by the user, which contributes to the motivation, understanding and learning of concepts that until then were abstract. It learns better what is understood by seeing.

> Virtual reality, by embodying, through image, sound or tactile sensation, concepts for which there is no physical model or representation, becomes a kind of translator, converting into concrete experiences that our senses can perceive, abstract ideas difficult to understand, facilitating learning and contributing to the development of cognitive abilities. (CAMACHO, 1996, p. 92).

In this way, the alliance between education and entertainment, which is often judged to be impossible, has been achieved. After all, teaching and entertaining are not opposing concepts, teaching does not need to be uninteresting, just as entertainment does not need to be devoid of educational elements.

5. TESTS AND RESULTS

In this chapter is the report of the product tests, carried out with undergraduates, postgraduates and professors of Higher Education. The tests consisted of the use of the simulator followed by the application of evaluation questionnaires.

The testing process of the simulator was carried out with graduates of the Biomedicine course of UEPA, Belém campus - PA, graduates of the Biological Sciences/Licentiature course of UFPA and IFPA, both in the Belém campus - PA; postgraduate students of PROFBIO of UFPA, Belém campus, and with six undergraduate teachers, who teach the discipline of Cellular Biology, one from UEPA, two from UFPA and three from IFPA. The results obtained considered the participant's observation during the simulator test and the questionnaires they answered.

5.1 Using the Simulator

The tests with the 92 undergraduates and the 6 postgraduates took place over 9 sessions. With the 22 graduate students of biomedicine from UEPA, 3 sessions were held, on different days and schedules, to meet their availability. The 6 sessions were held with students of Biological Sciences/Licentiature at the IFPA and UFPA during the course of Cellular and Molecular Biology (nomenclature of the course at the IFPA) and Cells and Molecules (nomenclature of the course at the UFPA). With the six post-graduates, it took only one session.

The course teachers used the simulator in the same session with students, some before starting the test with the students, others at the end of the test with them.

It was important to divide it into sessions, as the number of students in the classes in which the tests took place varies from 20 to 40. If everyone was present in a single session, individualized teaching would be impossible and experiments in VR would become difficult, for the time required, because of the need to occupy students who were not immersed in the virtual environment, because of the parallel conversations that arose. But in small groups while one student was immersed in the virtual world, the author managed to involve the other students in the activity, following on the computer screen (figure 64) what was happening in the virtual world. This made the exchange of opinions during and after the experience more efficient and enriching. The author was able to follow the immersion experiences in an individual way, from consulting to clearing doubts and to help (figure 65).

Figure 64 - Group interaction while a student is immersed

Source: author, 2019.

Figure 65 - Individual follow-up during immersion

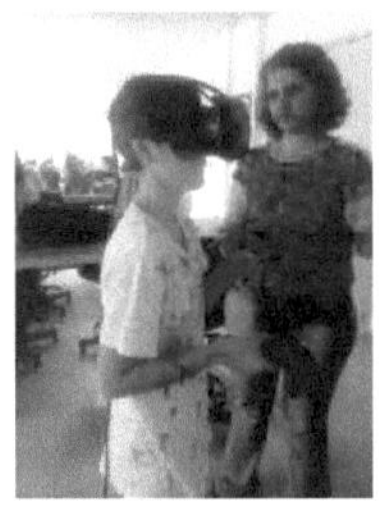

Source: author, 2019

The tests in the UEPA biomedicine course took place after the presentation of the product to the course coordinator, who showed interest and referred the researcher to the 2nd, 4th and 6th semester classes. The researcher explained the product and verified the interest of the class in participating in the testing process. As the interest was great, together with the students, the researcher planned three sessions on different days and times, according to the availability of the students. This would increase the quality of care and thus facilitate understanding of the subject.

In the courses of Biological Sciences/Licentiature at UFPA and IFPA, the researcher presented the product to teachers of Cellular and Molecular Biology (nomenclature at IFPA) and Cells and Molecules (nomenclature at UFPA). Again, there was great interest and they inserted the product test in the agendas of their subjects, as a practical activity.

In the UFPA course, 35 students from the 1st semester participated in the testing activities, which took place in two sessions that took place during the class time of the subject Cells and Molecules.

At the IFPA, two classes participated in the test, being one class of re-offering the subject Cellular and Molecular Biology, composed of 20 students, and the other class was from the 1st semester, with 15 students. The product tests in both classes took place in the class schedule of the subject, but in the freshman class, the individual immersion was more intense and detailed, so there was the need to hold a second session, so that everyone could participate in the testing process.

In the PROFBIO/UFPA post-graduation course, the tests took place after the researcher disclosed the developed product to the students themselves. A group of six students, who showed interest in participating in the test, determined the day and time they would be available for the session.

The application sites (laboratory, classroom) were organized so that the environments were free in all sessions, allowing users to move around in the real environment, which allowed a greater sense of immersion in the virtual environment.

At the beginning of each session, the researcher explained her motivation in developing the tool, how the simulator worked, the research objectives and how the product testing process would be (figure 66).

Figure 66 - Author explaining what the testing process would look like

Source: author, 2019.

For the testing process, the students received a Term of Free and Informed Consent (TCLE) (Appendix I) and, after filling out, each student used the *HTC Vive Oculus* and explored all virtual environment developed by the researcher. The objective was that they could evaluate the simulator individually, highlighting how was their experience and suggesting some improvements, if they thought necessary. And, during this process, it was possible to intuit that the students were able to achieve the sensation of total immersion of the virtual environment using the simulator, as can be observed in figure 67 (A) and (B); figure 68 (A) and (B); figure 69 (A) and (B); figure 70 (A) and (B); and figure 71 (A) and (B).

Figure 67 - Students exploring the virtual environment

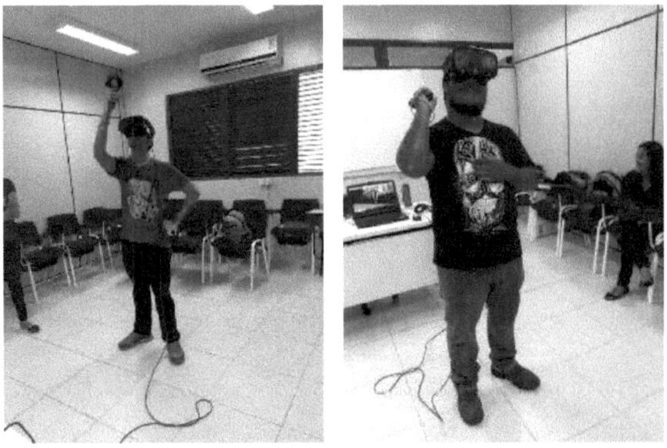

(A)   (B)

Source: author, 2019.

Figure 68 - Students exploring the virtual environment

 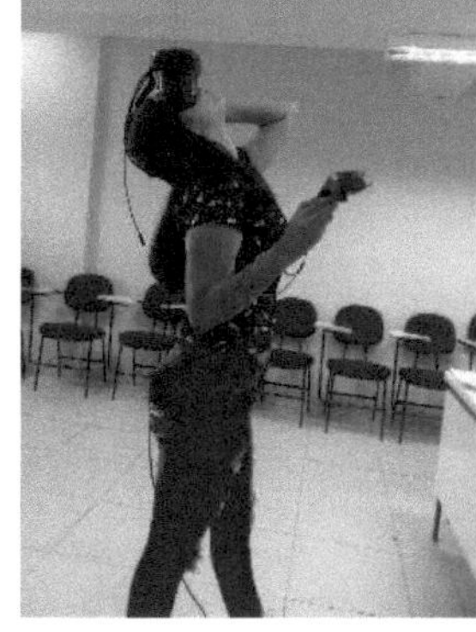

(A)   (B)

Source: author, 2019.

Figure 69 - Students exploring the virtual environment

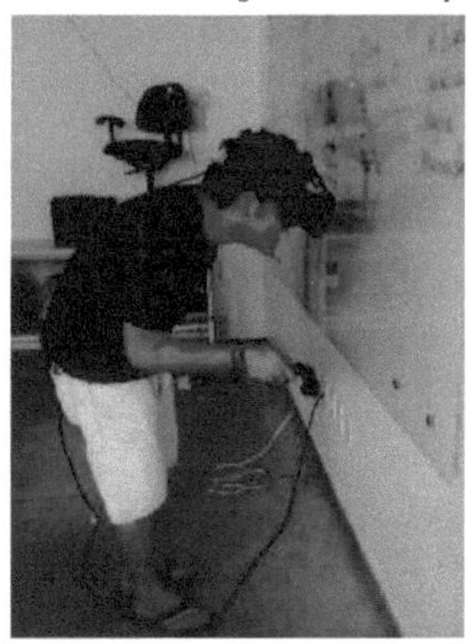 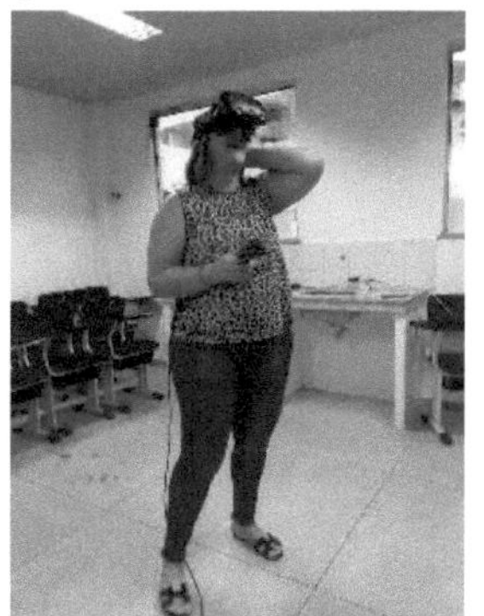

(A)   (B)

Source: author, 2019.

Figure 70 - Students exploring the virtual environment

 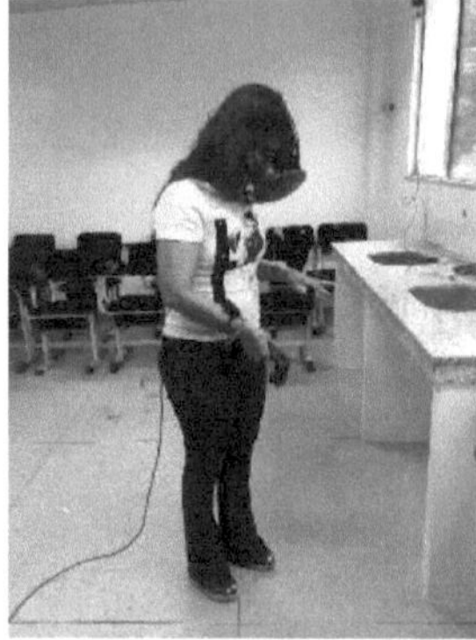

(A)   (B)

Source: author, 2019.

Figure 71 - Students exploring the virtual environment

(A)   (B)

Source: author, 2019.

This stage of experimentation was very satisfactory as, in general, the involvement of students during the use of the simulator was intense. They were enthusiastic about participating in the transport of substances through the membrane. During the interaction, some commented that the will was to stay in that environment to which it was inserted, because it was very fun to study that way.

5.2 Analysis of Experimentation Data with Students

At the end of the experiments, each student answered a questionnaire (Appendix II) to evaluate the simulator, considering aspects such as motivation during use, functionalities, contribution to the teaching learning process, among others.

The questionnaire consists of ten closed questions and seven open questions. In eight closed-ended questions, a Likert scale was chosen. This type of scale is made up of statements that are usually presented as a kind of rating table, in which statements are presented and the respondent is asked to give their degree of agreement with that statement, which varies from total disagreement to total agreement, including a moderate or neutral midpoint. The number of points that will make up the scale may vary between four and ten items, with the choice being at the discretion of the researcher. But it is important that the number of positive categories be equal to the negative ones, and you must not forget to include a neutral point (neither agree nor disagree; indifferent) (CUNHA, 2007; FRANKENTHAL, 2017; FIALHO, 2018). In this survey, five items were chosen (I totally disagree; indifferent; agree; totally agree).

As the VR simulator is not a frequently used tool in the teaching-learning process, it was deemed necessary to assess the difficulty of students in using the simulator. To do so, the following statement was made: "The simulator is difficult to manipulate". And, as can be seen in graph 1, most students did not have difficulties when using the simulator.

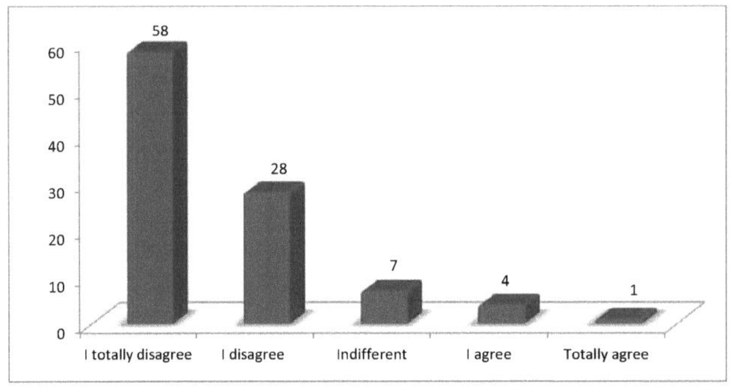

Graph 1 - The simulator is difficult to manipulate.

Such data confirm the observation of the researcher during the experimentation, when she noticed that, at the beginning of the immersion, some students had difficulty in handling the controller and moving in the virtual world, but, with little time of use and following her instructions on how to use it, soon they got used to the instrument and the interaction took place satisfactorily.

An advantage of the VR in relation to other interactive media, including the computer itself, is that, unlike the computer, whose use requires some minimal knowledge, which can inhibit its use to those who do not have this knowledge, out of "fear" and "shame" of making mistakes, the application of an VR system does not require the user any specific requirement related to its use. Even being based on a complex and sophisticated technology, it works in a simple way, not requiring knowledge of commands and keys, the participant just put the *HTC Vive Oculus* and use the controller (CAMACHO, 1996).

For the immersion to be efficient, it is important that the creator of the virtual environment provides clear information about the environment, so that the user has mastery of what is possible to do in the environment to which it was immersed. The key point to facilitate user interaction are the controllers, which need to be programmed in a way that facilitates user experience and immersion (MARTINS et al., 2017).

The researcher, when developing the virtual environment of the plasma membrane, tried to pass on the instructions for handling the simulator in the simplest and most objective way possible, so that the students would feel at ease during their immersion process in the virtual environment.

In order to know whether the information provided was sufficient for the user to perform well, the following statement was made: "The instructions provided were sufficient for use and handling of the resource". And, as seen in Figure 2, students generally agreed that the instructions provided were sufficient to use the resource satisfactorily.

Graph 2 - The instructions provided were sufficient for use and handling of the resource

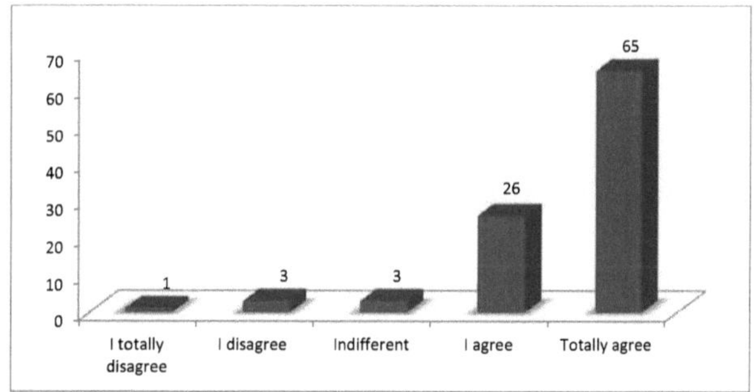

Source: author, 2019.

Education is an ongoing learning process that achieves better results when the person is motivated to learn and when there are adequate tools for this purpose. VR is a powerful tool to increase motivation and improve understanding of the most diverse subjects, especially those that are difficult to understand due to their abstract nature. The student can feel motivated to navigate and interact in the virtual world, becoming more interested in the subject being studied (RIZZATO; NUNES, s/a).

Seeking to verify if the simulator in a VR environment for teaching biology has achieved the benefits of interest, motivation and involvement of the student in the virtual environment, the following statements were made to students: "It was very interesting to use this resource"; "I enjoyed participating in an experience in Virtual Reality". It was found that most students considered immersion in the virtual environment to be interactive and enjoyed the experience (chart 3).

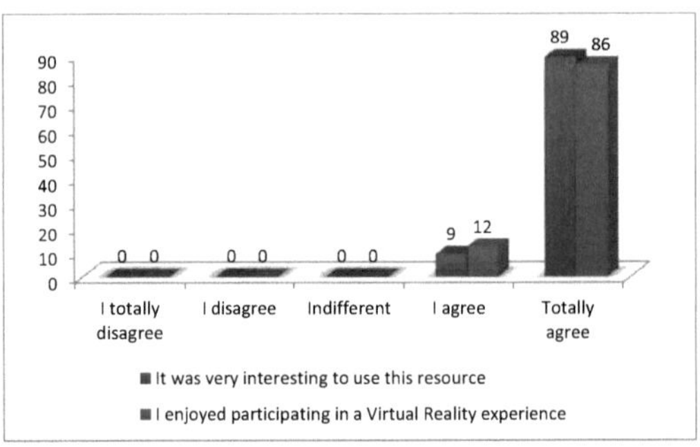

Chart 3 - Learner motivation and involvement when using the simulator

Source: author, 2019.

Students have different ways of understanding concepts and different approaches, in this sense, VR is considered a motivating pedagogical tool, because it acts as a learning object that contributes to the student not getting bored during classes, collaborates with self-study, respects the individual characteristics and learning pace of each student, which is fundamental in the teaching-learning process (BARILLI et al., 2011).

To see if the tool contributed to this motivation, the following statement was made: "At times I was bored while using the resource". And when analyzing the results, it was found that the tool was not classified as a boring educational resource (graph 4).

Graph 4 - At some times I was bored while using the resource

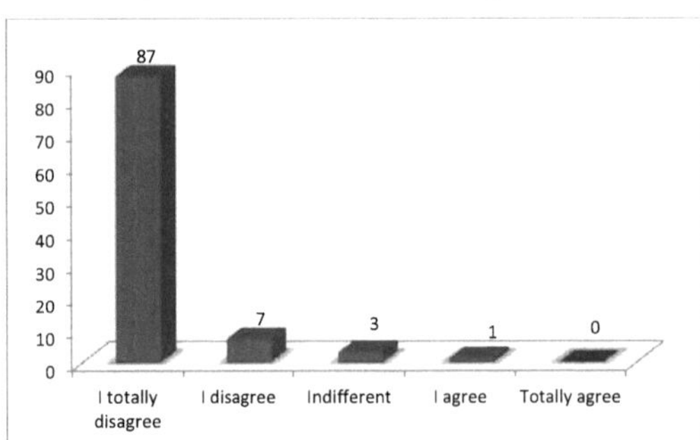

Although it is an excellent teaching tool, with numerous benefits offered, VR, like any other technology, does not replace the role of the teacher in the training of the student, and should not be pointed out as the solution to educational problems, or as something that will cause teaching methods to be completely changed. Like any other didactic resource, it should be used with defined objectives and under the guidance of the teacher, preventing it from being configured only as an entertainment tool, and acting positively in the teaching-learning process.

During the experimentation process, the researcher followed the immersion of each student individually and, simultaneously with this immersion, the students received an explanation from the researcher about the theoretical part of the content, when necessary. An interesting point for investigation is if the simulator developed by the researcher contributes to stimulate learning and assist in understanding the contents. Then, the following statement was made: "The use of virtual reality in teaching stimulates learning". And, as can be seen in graph 5, the students agreed that VR is configured as a stimulus in the teaching-learning process.

Graph 5 - The use of VR in teaching stimulates learning

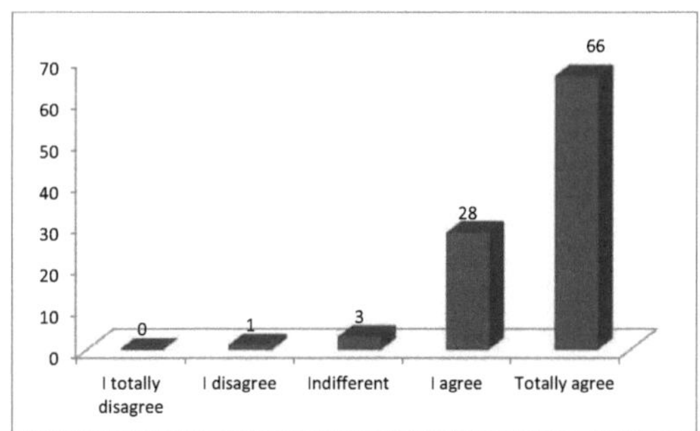

As for the contributions of the simulator in the teaching-learning process of the students, the researcher made the following statements: "After the application of the resource I understood the content better" and "I performed well while using the resource". And the students evaluated themselves positively, as can be seen in graph 6.

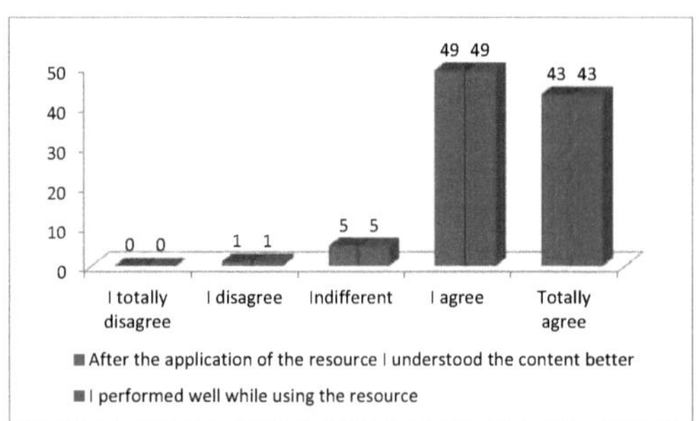

Chart 6 - The contributions of the simulator in the teaching learning process

Source: author, 2019.

The closed questions in the questionnaire ended with two questions in which the students were asked to mark "yes" or "no". They were asked if they would like to use the resources other times and if they would recommend the use of the simulator to other people. Only one student is not interested in using the simulator again. And the 98 students who participated in the testing process would recommend the use of the simulator to other people (Chart 7).

Chart 7 - Simulator evaluation and recommendation

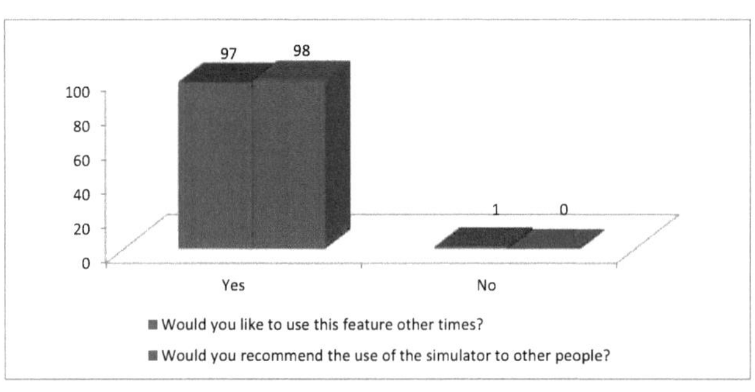

Source: author, 2019.

Based on the degree of agreement provided by the students and the analysis of the previous graphs, it can be seen that the product was well accepted and positively evaluated by the public investigated. Students find it easy to use; the information is clear and sufficient to handle; it is interesting to learn by means of VR, as in addition to stimulating learning, the resource facilitates understanding of the content. Therefore, this represents an indication of the positive potential of the simulator in VR environment for teaching the plasma membrane.

In the second part of the questionnaire, students answered seven open-ended questions so that they were free to express themselves about the question asked, most of them focused on questions specific to the content covered in the simulator.

The first question sought to gauge the difficulties in using the resource, so the following question was asked: "Did you experience difficulties using the simulator? Comment on your experience, highlighting what your difficulties were". In general, the students highlighted not having difficulties when handling the simulator, stating that its use was easy and intuitive. Those who claimed to have difficulties scored the handling of the controllers, however, stated that after the instructions of the researcher, the use became easy. Following, some answers from the students were highlighted:

"No. During that experience I felt no difficulty: it was completely safe. It brought me a greater understanding of the subject" (Student 1).

"No. I found the experience interesting and very enlightening. Being part of the cell allowed the understanding of the theory" (Student2).

"I had no trouble. I found the experience incredible, I felt inside the cell, as if I was part of it" (Student 3).

"I didn't have any difficulty using the simulator, I found it interesting to the point that I wanted to take everything that was happening inside the cell with my hand" (Student 4).

"There were no difficulties during the immersion. The controls are simple and very intuitive, allowing freedom of movement, making the visualization fluid and allowing different angles, ensuring better understanding of the processes visualized" (Student 5).

"No, I found it extremely interesting to use this resource, because it was possible to observe more clearly some movements that occur in the phospholipidic layer and others. I had no difficulty in using the resource or in understanding the content" (Student 6).

"No. The guidelines were very clear and objective" (Student 7).

"Not much, since the instructions are very didactic" (Student 8).

"Initially a little difficulty with handling the control, but the main difficulty was a little dizziness during the virtual environment" (Student 9).

"Very little, the simulator brings us a reality that is not seen in the classroom, this makes learning very didactic" (Student 10).

"I felt a bit difficult at first, because I felt a strange sensation at the beginning of the experience, but after I "dived" into the virtual experience, it became more comfortable and I was able to enjoy the environment. At first I was tense" (Student11).

"The difficulty I felt was in relation to the ergonomics of the virtual glasses, because I'm short-sighted and it bothered me a bit to superimpose the glasses, but it didn't totally hinder the experience" (Student 12).

"I felt little difficulty, it was a very interesting experience, where I did not have much difficulty because of the continuous orientation" (Student 13).

"Yes, my difficulties were with handling, questions like adapting to commands, knowing where the buttons are, but they were alleviated as I used the product" (Student 14).

"I found it difficult to understand the processes I was visualizing, but they became easier with the developer's instruction" (Student 18).

Analyzing the answers of the students, it is proved what MARTINS et al. (2017) show regarding the programming of controllers. According to the authors, it needs to be done in a way that facilitates the user's experience and immersion.

The answers also prove that VR does not replace the presence of the teacher, on the contrary: it is a tool that helps the teacher in teaching, bringing practice to the student, but before that, it is necessary to have knowledge of the theory. From the statements before highlighted, it is possible to see that the doubts and difficulties of the students were resolved as the researcher intervened in the process of immersion.

The second question aimed at knowing what the student understood best after using the simulator. As described in section IV, the developed simulator demonstrates all the structures that constitute the plasma membrane, the movements of phospholipids and the transport of small molecules through the plasma membrane. Among these transports, the following are part of the simulator: simple diffusion; facilitated diffusion; ionic channel; selectivity of the sodium channel; selectivity of the potassium channel and the sodium and potassium pump. Of these subjects, the most understood by the students were the movements of phospholipids; the structure of the plasma membrane and simple diffusion, as shown in graph 8.

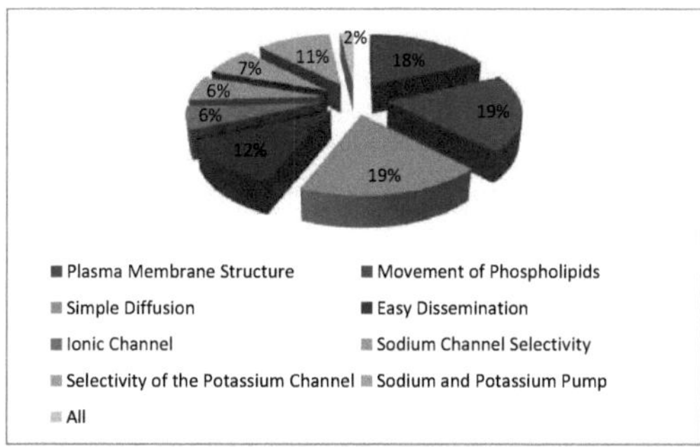

Source: author, 2019.

It was thought interesting to highlight some answers, as they enhance the importance of using didactic resources that bring the student closer to practice:

"Everyone! Because I have visual memory and in the simulator I could see all the constitution of the membrane and perform the transports made by it" (Student 2).

"Finally, now in the 6th semester of the course, I was able to understand the real difference between simple dissemination and facilitated dissemination" (Student 3).

"All, for I have been able to observe the contents that in the classroom are so theoretical and abstract. With the simulator, I had the opportunity to feel inside the cell, that connected me with the subjects" (Student 15).

"The sodium and potassium pump, only today, after the use of the simulator, it became clear to me how the ATP participates during this process" (Student 16).

"Everyone! It was very innovative and interesting to rescue in this way content that I learned years ago, when I was still in the first semester. If at that time the teacher had made use of that refusal, the class would have been much more interesting" (Student 17).

One of the basic concepts of the plasma membrane, highlighted in Chapter I, is that the plasma membrane is a structure consisting basically of a bi-layer of phospholipids with proteins inserted in this layer. As a study model, the so-called fluid mosaic model is used to show that the membrane is fluid, since its components are capable of moving through the structure, and is therefore not a completely static structure. Proteins and lipids have the ability to move.

To assess whether the students were able to visualize this during their immersion in the plasma membrane, the author made the following question: "The basic structure of all cell membranes is a phospholipidic layer. During the use of the simulator, what were the movements performed by these structures that you were able to visualize?". In the answers of the students, it was noted that even though the movements of phospholipids are evidenced in a virtual environment, built exclusively to visualize the movements of these structures, some of their movements were not visualized by the students (graph 9), which is understandable because some movements are very specific in the hydrocarbon tail, requiring the student to "dive" into the phospholipid layer. And the user's immersion in the environment is a personal experience, so even the teacher/researcher advising that certain movements would only be visualised if the student entered the phospholipidic layer, some chose to visualise from afar.

Graph 9 - Movements of Phospholipids visualized during immersion

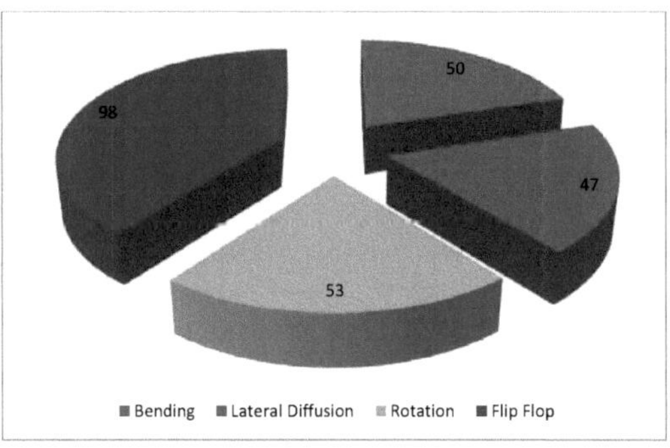

Source: author, 2019.

Analyzing the graph, it is possible to conclude that only the "flip-flop" movement was visualized by all the students who participated in the validation process. And the lateral diffusion movement was the one that had the least visualization.

Two students not only cited the movements they were able to visualize, but also reported their experience in seeing these movements:

"I visualized every move. The fact that they are visible has allowed a better understanding of how they occur. I found it interesting that they moved in all environments, I

104

saw them moving even while visualizing the environments of transport of substances" (Student 11).

"Flip flop, rotation, flexion and lateral diffusion. All were very well illustrated and the fact of having the possibility to enter the middle of the plasma membrane and follow the movements closely, was an incredible experience" (Student 13).

The proteins that make up the plasma membrane are well differentiated. Having as reference Alberts et al. (2017), figure 22, present on page 67 of this research, the author modelled eight different types of proteins and inserted them in the didactic plasma membrane model, figure 29, page 71. In order to quantify how many of these proteins the student identified, the following question was asked: "What types of membrane proteins did you observe during the use of the simulator? Exemplify and cite differences between them". Of the ninety-eight questionnaires analyzed, the most frequently cited proteins were the transmembrane proteins α- single helix, α multiple helix and the protein β. Only six students cited peripheral proteins and in this questioning there were eight blank answers. 25 questionnaires presented the exemplification, all based on protein morphology. In four of them, the student had difficulty in using the specific biological terms to exemplify it and they chose to make schematic drawings (figure 72), representing the proteins they visualized in the simulator.

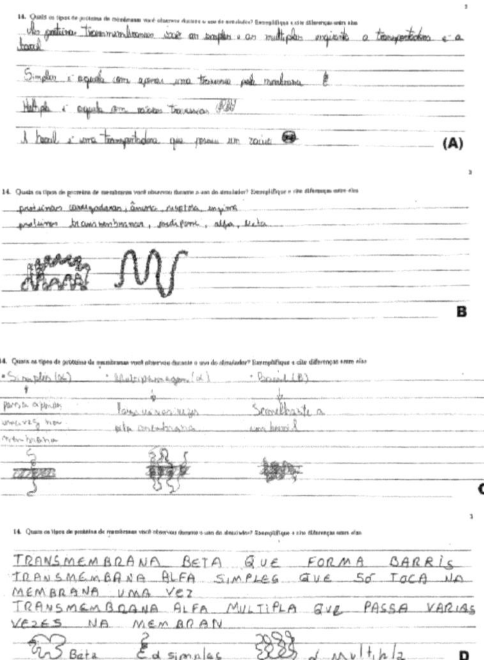

(A) Illustration Student 14; (B) Illustration Student 28; (C) Illustration Student 32; (D) Illustration Student 63.
Source: author, 2019.

Among the eight virtual environments present in the *menu*, seven represent transport of substances through the plasma membrane. In order to explore the students' theoretical knowledge about these transports, the author made the following question: "What types of transport did you notice during the use of the simulator? Exemplify". Three students did not answer this question. The most commonly cited transports were simple diffusion, easy diffusion and sodium and potassium pumps. As for the example, only 31 students exemplified. And among these examples, four errors were found in the theoretical concepts.

"Active transport: directly through the membrane. Passive transport: passes through a channel to facilitate" (Student 17).

"Active and passive. In active, the passage is selective. On the passive, the selective does not occur" (Student 18).

"Simple or passive transport that a molecule enters the membrane without spending energy and the facilitated diffusion that is very selective and only occurs with energy spending of the cell, during the process" (Student 19)

"Facilitated needs the ATP to work, passing the ions. Simple, does not need to be ATP spent" (Student 20)

The example of student 17 is wrong, because in active transport, substances pass through a transmembrane and only occur in the presence of energy. In simple passive transport, the substance passes through the phospholipidic layer, and if the passive transport is facilitated, it passes with the aid of a protein. Passive transport occurs without energy expenditure. Student 18's error lies in the fact that it does not matter if the transport is passive or active, the passage is always selective, since this is one of the functions of the plasma membrane: to select the substances that enter and leave the cell. In the examples of students 19 and 20, the error lies in stating that, in the facilitated diffusion, there is expenditure of energy.

Ion channels are very selective as discussed by the author in section II, and the highest ion selectivity index is given to the Na+ and K+ channels, both represented in the simulator. To verify if the student understood this ion selectivity, the following question was asked: "Ion channels are highly selective, allowing the passage of only ions with appropriate size and loads. Did the simulator allow you to observe this? Exemplify". The analysis of the answers was very satisfactory because, of the 90 answers analyzed, all presented the example (eight students left this answer blank), therefore, one can evaluate the tool positively, because it made real such an abstract content, which allowed a better understanding of the content.

Seeking to make the simulator the most didactic and understandable, the questionnaire closed with the following question: "In your opinion, what could be improved in the virtual environment so that your experience and learning are more positive? The following are some answers to this question:

"The environment is already very explanatory and dynamic, I believe that by adding other subjects in the future, it can help students and teachers in other subjects" (Student 20).

"If possible, allow the student to pause the movement of the transport of substances at the time he or she sees fit" (Student 21).

"Around the membrane, the environment surrounding the membrane, rather than a room, could be a human organism, showing several other cells around" (Student 22).

"It would be nice to take this method to other disciplines, because it was very productive" (Student 23).

"The colors are very strong, long time looking at the colors green, yellow, red starts to hurt the vision" (Aluo 24)

"Instead of occurring in a room, it could be within an individual" (Student 25).

"I found it very positive and creative, because we have the possibility of leaving the abstract membrane for the real world of the membrane. Suddenly for future work show endocytosis and exocytosis block transport" (Student 26)

"Nothing, I found highly didactic and meaningful to learning" (Student 27).

"More simulators to be used for longer" (Student 28).

"Nothing. The simulator is very well developed didactically" (Student 29)

"I really liked the colors, but if the files had subtitles, the experience could be more independent." (Student 30).

"Standardize colors where even colorblind can differentiate the indicated specificities and allow more dynamic activities in the virtual world" (Student 31).

"I could have a voice instructor, the handling would be more independent. But this educational technology was superbly developed and the best thing about it is that it can be applied to other contents" (Student 32).

Taking into account the responses, 50 students who considered the simulator satisfactory and that it did not need any modification, and those who suggested change, highlighted the following improvements: audio narrating what was happening in the virtual environment; presence of an *avatar* in the virtual environment explaining the processes; subtitling of structures in all environments; change in colors, because they are strong and bright; in the cell environment detail its other components (cytoplasmic organelles and nucleus); possibility of having more *HTC Vive Oculus;* change of the environment for an organism; possibility of pausing transports; making simulators like this for other contents; correction of "defects" of duplicate images. These suggestions were accounted in table 1.

Table 1 - Suggested changes to the simulator

SUGGESTIONS FOR CHANGE	NUMBER OF STUDENTS WHO SUGGESTED

Audio narrating what was happening in the virtual environment	05
Presence of an *avatar* in the virtual environment explaining the processes	02
Legend of structures in all environments	18
Change in colors, because they are strong and bright	06
In the cell environment detail its other components (cytoplasmic organelles and nucleus)	09
Possibility of having more *HTC Vive Oculus*	02
Changing the environment for an organism	02
Possibility of pausing transport	01
Make simulators like this for other contents	16
Correction of "defects" in duplicate images	12

Source: author, 2019.

5.3 Analysis of Experimentation Data with Teachers

Besides the students' testing, it is important for the author a *feedback* from teachers who teach the subject of Cellular and Molecular Biology. The teachers of the classes who participated in the testing process were themselves invited to use and evaluate the product.

In UEPA, two teachers were informed by the coordinator of the biomedicine course that the students were participating in the testing of a virtual reality simulator aimed at teaching the plasma membrane. These teachers became interested and attended the testing session to use the resource. At the IFPA, besides the teachers of the classes that participated in the testing process, another Cell Biology teacher became interested in the resource and asked to use it.

Six teachers participated in this testing process, two from UEPA, three from IFPA and one from UFPA. The evaluation process was also by questionnaire (Appendix III), consisting of seven closed questions using the Likert scale, two closed questions with the affirmative/negative options and six open questions for them to express themselves about the tested product.

In order to verify if the general objective of creating the simulator described in this research was achieved, the following statements were made: "Using this simulator can make the teaching-learning process of the subject more interesting", "It is easier to teach Cell Biology using this simulator or others similar to it". And all teachers agreed with the statement (graph 10).

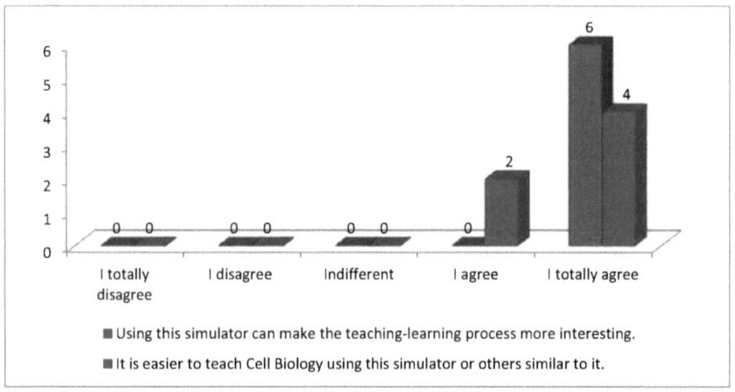

Source: author, 2019.

The use of the simulator does not replace the theoretical lesson, it was developed to facilitate learning, but, by itself, does not solve the problem of teaching the plasmatic membrane, becoming much more efficient when the student has already had contact with the theory. To verify the position of the teachers in relation to this, the statement was made: "The simulator can be used to review the subject worked on". And everyone totally agreed with the statement.

In order to check whether the simulator matches the profile of its target audience, the following statements were considered: "The information presented is useful, objective and contributes to understanding the subject matter", "The simulator is easy to handle", "The representations of the interface information are easy to recognize (*menu*, modeling etc.)". And, when analyzing the result (graph 11), one notices that the product is adequate.

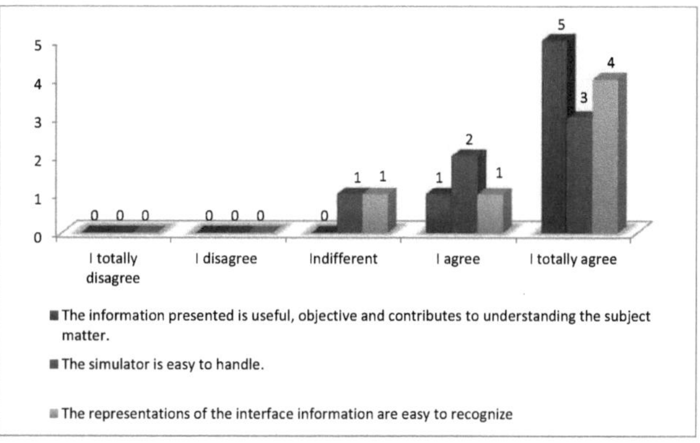

Graph 11 - Simulator content and interface

Source: author, 2019.

The last statement of the questionnaire was: "It would be interesting to have simulators like this addressing other contents". And everyone agreed completely, which shows that the resource was evaluated in a positive way, to the point that teachers thought about the possibility of adopting VR to address other issues.

After the bibliographic research presented in section III of this research, it was found that there is a shortage of educational products using VR in Brazil. In order to verify if the teachers knew this technology as a teaching tool, the following questions were asked: "Have you ever had contact with any virtual reality resource focused on teaching? ( ) Yes   ( ) No. If the answer to the previous question is yes, describe where and how the experience was". Everyone answered that they had never used teaching-oriented VR.

When asked if they believe that VR is a good resource to be used in the teaching of Cellular and Molecular Biology, the following answers were obtained:

"Yes. It is an excellent resource because it allows the visualization of the structures of the plasmatic membrane and characterizes its functions, making less abstract the teaching of this subject in cellular and molecular biology" (Professor 1).

"Yes, mainly because of the possibility of awakening greater interest in the teaching-learning process" (Professor 2).

"Yes. It's an exceptional resource, I've never had the chance to use it before, and I was completely enchanted by the feeling that it's inside a cell. Apart from the possibility to visualize and experience such abstract contents" (Professor 5).

"Yes. The resource is fantastic and extremely necessary, the content taught in Cellular and Molecular Biology for higher education is difficult to assimilate, and there is a need to use new and current methodologies and practices for their assimilation. I believe that with virtual reality, the student will experience a new approach to content, making it attractive, interesting and, consequently, will facilitate their learning. The tool becomes fundamental, because the discipline is the basis of the course and serves as a prerequisite for other disciplines of specific content, being the basis for training the professional in the area" (Professor 3).

"Yes. The content of Cell Biology is significantly abstract, making it difficult for students to understand. By using virtual reality glasses, it is possible to enable students to interact with the content that may result in better assimilation of the content" (Professor 4).

"Yes. In twenty-five years of minsitrating Cell and Molecular Biology I had never seen a didactic resource that would make the discipline less abstract. Today I have been on an exodus by using this didactic resource and I intend, if possible, to use it in my classes as soon as possible" (Professor 6).

They were then asked if they would adopt the simulator in their Cellular and Molecular Biology classes, justifying their answer. To the researcher's satisfaction, they all stated that they would adopt it. Below, the justifications presented:

"Yes, because it would allow us to demonstrate the structure and functioning of the membrane in a more concrete and interactive way, which would certainly facilitate the understanding by the students" (Professor 1).

"Yes, mainly because of the possibility of awakening greater interest in the teaching-learning process" (Professor 2).

"Yes. I believe the simulator would contribute to a better assimilation of the content" (Professor 4).

"Yes. Because it will allow students to realise such abstract and complex content, so I believe there is an improvement in their learning" (Professor 5).

"Absolutely, because it feels great to be inside the cell" (Professor 6).

"Surely any didactic resource that escapes from traditional methods and is proven to influence the assimilation and learning of such specific content should be used and explored. Virtual reality allows a new universe to the student, as it allows him to enter the cellular universe and manipulate the cellular components through the resource. Besides what

accompanies the evolution of technology being essential for the century we live" (Professor *III*).

To identify possible flaws and/or errors in the simulator, three questions were asked: "Are the concepts of molecular and cellular biology addressed in the simulator correct?"; "Regarding the design of the simulator, do you consider that the formats and colors used in modeling are appropriate? ( ) Yes (   ) No. If not, what could be improved?"; "Do you consider that any aspect of the content covered in the simulator needs to be appropriate and/or improved? What?".

No conceptual errors were pointed out by the teachers, as well as no suggestions for modifying the modeling of cell structures. Regarding the adequacy and/or improvement of the content approached in the simulator, four said that the content is good and appropriate to what was proposed; one teacher scored having nothing specific to score because he did not use the simulator for that purpose. But he finds it interesting that later he has the opportunity to do that, because everything "can always be improved". The suggestions given were: the addition of a bookcase in the museum, to present a glossary or summary of the main concepts, description of structures and dynamic events of the membrane; and that the researcher would continue to develop, later, other functionalities of the membrane and even other contents of Cellular and Molecular Biology.

5.4 Final Test Considerations

After testing the product and comparing the data obtained, it was possible to observe that the simulator had a good acceptance by students and achieved its goal of facilitating the learning process, to the point that some students suggest simulators such as this by addressing other content.

As for the suggested changes, the colors have already been changed to a more matte tone. The subtitles have not been placed, as the researcher considers that there will be a visual pollution due to the large amount of information, but plans to make a tutorial for each virtual environment, similar to what appears at the beginning of the immersion. The objective of developing these new tutorials is to summarize the theoretical approach of each subject approached in the simulator.

Regarding the "defects" cited by the students, the author already expected such scores, however, this is not a simulator error, but problems in the installation of *HTC Vive Oculus*, since not all environments provided for product validation met the installation specifications explained by the author in chapter III, in the section on integration into the virtual environment, figure 41, page 81. This caused problems in sensor communication, allowing the duplication of images during the immersion, being necessary to restart the equipment several times.

Analyzing the teachers' responses, it is possible to ratify what was discussed in the first chapter of this work, about how much Cellular and Molecular Biology presents abstract and difficult to understand contents, being necessary that teachers adopt different methodologies. Through the teacher's evaluation carried out in the testing process, it is possible to see that the simulator in VR environment for teaching the plasma membrane is configured as a pedagogical resource enhancing the teaching-learning process.

6. COMPLETION AND PROPOSALS FOR FUTURE WORK

Science is experience, and experience is a cloak that is woven over several centuries; and the more the cloak extends, the more science is complete and secure.

(C. Bini)

The last chapter of this dissertation aims to establish conclusions about the work and to present proposals for the development of future work related to the main theme.

In the course of the research, it was found that the main challenge for teachers is to seek methodologies that lead to learning and not just teaching. And for such methodologies to work, the student needs to go beyond the passive role of just listening, reading, memorizing, faithfully repeating the teacher's teachings and becoming creative, critical, researcher and active to produce knowledge.

Considering these aspects, a simulator has been developed in VR environment for the teaching of plasma membrane, in which teachers and students not only have a three-dimensional view of cellular structures, but also can manipulate these structures and simulate the transport of small molecules, with complete interactivity, thus leaving the field of imagination.

With the proposed simulator, the student does not have contact with the content only through the teacher's speech or through images with which he cannot interact, but through his own experience and experimentation, ceasing to be passive in his learning process, because he will not learn from a description made by another person, as occurs in the expository classes, which provide an objective, conscious and implicit learning. When immersed in the plasma membrane, the student becomes active in his/her learning process, as he/she knows the membrane from his/her interaction with it, as detailed in chapter III.

During the course of the construction of the tool, it was necessary to carry out research to support the use of VR in teaching. These researches were exhaustive, as most of the articles found, based on previously defined keywords, were not related to the proposal of this research. Even being a tiring work, due to the scarcity of results obtained, it was decided to repeat the research in the months of September and October 2019, which allowed the enrichment of the work, because two articles published in 2018 and five published in the first semester of 2019 were found, which related the content of Cell Biology with the VR.

Among the major difficulties faced during the development of the simulator, the interdisciplinary character of the tool stands out, which forced the author to leave her comfort zone and explore new contexts. Because its construction not only required the biological concepts of the plasmatic membrane (the author's training area), but also modeling and programming knowledge, which was the great challenge in the development of this product.

It was necessary, then, the formation of an interdisciplinary team, which emerged with the partnership of the author as LabITE and LAAI, to assist this development process. The first step for the product development was to learn how to model the structures that make up the plasma membrane in *Blender software*, a process that lasted approximately three months. The learning took place through tutorials on the *Internet* (APOSTILABLENDER; BLENDER280; BLENDERTUTORIAIS; B., 2013; BORGES, 2018; DASSIE, 2015; MONQUEIRO, 2011) that taught the process of modeling and classes with a LabITE *design* professional. For not being used to the area of Computer Science, the modeling process was very slow, some cellular structures were only possible to be modeled after accompanying a professional in the area modeling, to then be able to make their own modeling.

After the development of the simulator was completed, the testing process was carried out, which considered several aspects, such as the content addressed in the simulator, which, according to the data analyzed above, were positively evaluated by teachers and students. There are reports that the simulator contributed significantly to the teaching learning process, such as, for example, a student in the sixth semester of Biomedicine, who stated that only when he used the simulator was he able to understand the difference between simple diffusion and facilitated diffusion. This highlights the importance of making "real" a content that is so abstract.

Before analyzing the data from the testing process, it is possible to conclude that the simulator achieved its general objective and the specific ones, because it was considered as a teaching tool that contributes to facilitate the learning of the plasma membrane content, besides making this learning playful and enjoyable, being also pointed out the viability of other contents and/or disciplines to use this type of technology.

It was very satisfactory to note that the didactic resource proposed enabled an innovation in the presentation of the content of the plasma membrane. And it made a difference in the teaching-learning process.

6.1 Future Works

The conclusion of a research is not configured at the end of a work, because, during the realization of this work, new worries and ideas arise for the realization of new activities, but many times they are not developed because of the deadline for the end of a dissertation.

During the development of the simulator, the researcher planned the construction of a puzzle of phospholipid models presented in this dissertation, because she believes that, through this didactic modeling, it will create opportunities to improve learning, allowing the student to have experiences that awaken their motivation for inquiry and for investigative attitudes, in which they seek solutions.

The current moment demands new teaching practices, so that educators assume a role as catalysts of the creative potential of each student, which would certainly result in better use of talent and human potential in the educational context (ALENCAR, 2002).

Seeking to innovate the teaching-learning process of the biochemistry discipline, the goal of building the phospholipids puzzle is to propose a didactic strategy based on 3D printing of phospholipids for a better visualization and learning of their molecular structure.

The puzzle being designed will consist of atoms (carbon, hydrogen, oxygen, nitrogen and phosphorus), which will be printed on the 3D printer; pieces of wire and a guide to guide the construction of three-dimensional models of phospholipids (Appendix IV). The process of building this puzzle has already begun, and in the first 3D models developed, the atoms were pierced with the amount of chemical bonds they establish, however, the researcher concluded that it would be very intuitive and would not allow the educator to observe if the students understood the theory. Thus, the atoms were remodeled with various perforations. That is, for the construction of 3D models (figure 73), the students will need scientific knowledge of the amount of chemical bonds established by each atom. Such bonds will be represented by wires, as they present greater resistance when compared to the rods designed and developed in the 3D printer.

Source: author, 2019.

During the assembly of the pieces, the student will be able to see what he is learning, becoming more active in the teaching-learning process and facilitating the memorization of the contents. This is because, when assembling the pieces, he will be able to visualize and understand the chemical bonds established by the atoms that constitute the phospholipids.

The researcher believes that this puzzle is important for the learning process because it materializes an abstract knowledge. The student will be able to assemble the spatial formulas of the five types of phospholipids that make up the plasma membrane, which will allow greater participation of these during the classes, in addition to a greater understanding of how these atoms are organized when establishing chemical bonds.

Besides the construction of the puzzle, considering the suggestions provided during the testing process, the virtual environment of the plasma membrane will be improved. The theoretical description of what occurs in each virtual environment is already being developed. The process of transporting substances through the plasma membrane will be expanded to explore the passage of large molecules through the processes of endocytosis and exocytosis.

The suggestions to explore the entire cellular environment, instead of focusing only on the plasma membrane, were accepted. Then, all the cytoplasmic organelles will be properly identified and there will be a *menu* presenting the functions of each one. The same will be done with the structures that compose the cell nucleus.

To facilitate access to the developed product, the simulator is being suitable for use in *smartphones* and explored through models of RV glasses that suit this media.

As the product developed presents a new way to approach such complex contents, the perception of interest and motivation of the students and teachers who used the tool was remarkable. And as they themselves suggested, it is interesting to think of new virtual environments for other Biology contents, such as microbiology, virology, biochemistry.

BIBLIOGRAPHICAL REFERENCES

ALBERTS, Bruce et al. Bioenergetics and cellular chemistry. In: ______. **Cell molecular biology**. Trad. Andrane Ardala Elisa Breda et al. Porto Alegre: Artmed, 2017.

ALENCAR, E. M. L. S. The educational context and infuence in creativity. **Critical Lines**, Vol. 8, pp. 165-205, 2002. - 15.

ANDRADE, Julia Pinheiro; SARTORI, Juliana. The author teacher and significant experiences in 21st century education: active strategies based on the methodology of contextualization of learning. In: MORAN, José Manuel; BACICH, Lilian Cássia. **Active methodologies for an innovative education**: a theoretical and practical approach. Porto Alegre: Penso, 2018.

Apostille Blender. Available at: http://coenc.td.utfpr.edu.br/. - January 8, 2018. - http://coenc.td.utfpr.edu.br/~giron/HIPER/Blender.pdf.

B. Jeremy. **Becoming a professional in Blender 3D.** Available at: https://pt.wikibooks.org/wiki. - May 3, 2013. - November 28, 2017. - https://pt.wikibooks.org/wiki/Tornando-se_profissional_em_Blender_3D.

BARILLI, Eloma Christina Vieira Castilho; EBECKEN, Nelson Francisco Favilla; CUNHA, Gerson Gomes. Virtual reality technology as a resource for distance public health training: an application for learning anthropometric procedures. **Collective Health Science**, 1: Vol. 16, pp. 1247-1256, 2011.

BARRET, Kim E. et al. Cellular and molecular bases of medical physiology. In: ______. **Ganong's Medical Physiology**. Porto Alegre: AMGH, 2014.

BAZZO, V.L. Where do the degrees go? Teacher training and public policy. **Education**, 1. Vol. 25, pp. 53-65, jan/jun 2000.

BEHEIRY, Mohamed El et al. Virtual Reality: Beyond Visualization. **Journal of Molecular Biology** / ed. reserved Elsevier Ltd. All rights. - 7. Vol. 431, pp. 1315-1321, março de 2019.

BEHRENS, Marilda Aparecida. Collaborative Learning Projects in an Emerging Paradigm. In: MORAN, José Manuel; MASSETO, Marcos T.; BEHRENS, Marilda Aparecida. **New Technologies and Pedagogical Medication**. Campinas: Papirus, 2013. - 21.

BENNET, Jennifer A.; SAUNDERS, Colin P. A Virtual Tour of the Cell: Impact of Virtual Reality on Student Learning and Engagement in the STEM classroom. **Journal of Microbiology & Biology Education**, - 2. Vol. 20, 2019.

BLENDER 2.80 REFERENCE MANUAL. Available: https://docs.blender.org. - December 15, 2017. - https://docs.blender.org/manual/pt/latest/.

Blender Inc. Initial Blender website. Available at: www.blender.org. - June 15, 2018. - March 8, 2019.

BORGES, William. **How to model in BLENDER well explained**. Very easy. Available at: https://youtu.be/8IDtkj7eWJo. - January 8, 2018. - January 12, 2018. - https://youtu.be/8IDtkj7eWJo.

BRAGA, Mariluci. Virtual Reality and Education. **Journal of Biology and Earth Sciences**, - 1: Vol. 1, 2001.

BRITO, Allan. **Blender 3D**: interactive games and animations. São Paulo: Novatec, 2011.

CADOZ, Claude. **Virtual Reality**. Trad. GOYA Paulo. São Paulo: Ática Publishing House, 1997.

CALZA, R. E.; MEADE, J. T. The GenTechnique Project: Developing an Open Environment for Learning Mollecular Genetics. **Computers & Education.** - [s.l.]: Elsevier Science Ltd All rights reserved, - 1-2 : Vol. 30, jan/fev de 1998.

CAMACHO, Lurdes. **Memories of a Future Time**: virtual reality and education. - [s.l.]: Hugin - editores Lda, 1996. - 1: p. 345.

CARDOSO, Alexander; LAMOUNIER, Edgard. The Virtual Reality in Education and Training. In: TORI, Romero; KIRNER, Cláudio; SISCOUTTO, Robson. **Fundamentals and Technology of Virtual and Augmented Reality**. Porto Alegre: SBC Publishing Company, 2006.

CARVALHO, Hesli de Araújo. **Virtual Reality in Education**: a study of the Brazilian situation: Monograph. Lavras: [n.º], p. 71, 2002.

COOPER, Geoffrey M.; HAUSMAN, Robert E. **The cell**: a molecular approach. Porto Alegre: Artmed, 2007. - 3.

CUNHA, Luisa Margarida Antunes. **Models and Scales of Likert and Thurstone in measuring attitudes**: Dissertation / University of Lisbon. Lisbon: [n.º], 2007.

DASSIE, Marcos. **Tutorial**: Introduction to 3D modeling with Blender = Tutorial: Introduction to 3D modeling with Blende. Available at: https://www.fabricadejogos.net/. - October 7, 2015. - October 12, 2017. - https://www.fabricadejogos.net/posts/tutorial-introducao-modelagem-3d-com-blender/.

DOLAN, E. L.; COLLINS, J. P. We must tech more effectively: here are four ways to get started. **Molecular Biology**, - 12: Vol. 26, 2015.

EICHLER, Marcelo Leandro; PINO, José Cláudio Del. **Virtual Learning Environments development and evaluation of a project in environmental education.** Porto Alegre: Publisher of UFRGS, 2006.

FERNANDES, Cláudio Henrique de Souza; ZAMA, Uyrá dos Santos. **Biomembranes and teaching by research in higher education**: Dissertation / Federal University of Ouro Preto. Ouro Preto: n.], 2018.

FIALHO, Arivelto Bustamante. **Virtual and Augmented Reality**: technologies for professional applications. São Paulo: Ethics, 2018, - 1: p. 144.

FIOLHAIS Carlos; TRINDADE Jorge Fonseca. The virtual reality in the teaching and learning of Physics and Chemistry. Physics **Gazette**. Lisbon: Portuguese Physics Society, - 19: Vol. 2, pp. 11-15, 1996.

FIOLHAIS, Carlos; TRINDADE, Jorge Fonseca. Physics for All - Misconceptions in Mechanics and Computational Strategies. **COLLOQUIUM OF PHYSICS OF THE POLYTECHNIC INSTITUTE TO TAKE.** Tomar: Instituto Politécnico, 1999, pp. 185-202.

FRANKENTHAL, Rafaela. MindMiners Blog. **Understand the Likert scale and how to apply it in your research**. - May 24th, 2017. - July 9, 2019. Available at: https://mindminers.com/blog/entenda-o-que-e-escala-likert/.

GALDINO, Any Karolyne. engenhariae.com.br. **This is how Virtual Reality was born in 1968**. - February 28, 2019. - October 5, 2019. Available at: https://engenhariae.com.br/editorial/colunas/e-assim-que-a-realidade-virtual-nasceu-em-1968.

GARCIA-BONETE, Maria José; JENSEN, Maja; KATONA, Gergely. A practical guide to developing virtual and augmented reality exercises for teching structural biology. **Biochimestry and molecular Biology Education.** - [s.l.]: International Union of Biochemistry and Molecular Biology, - 1 : Vol. 47. - pp. 16-24, jan/fev de 2019.

GLASSER, William. ppd.net.br/william-glasser. PPD - **DYNAMIC PEDAGOGICAL PROJECTS.** - 2017. - May 21, 2019. Available at: http://www.ppd.net.br/william-glasser/.

GOODSELL, David S.; FRANZEN, Margaret A.; HERMAN, Tim. From Atoms to Cells: Using Mesoscale Landscapes to Construct Visual Narratives. **Journal of Molecular Biology.** - [s.l.]: Elsevier Ltd. All rights reserved, - 21, - Vol. 430, pp. 3954-3968, 21 de outubro de 2018.

HARTSHORN, Michael J.; HERZYK, Pawel; HUBBARD, Roderick E. Exploring molecular structure by virtual reality. **Trends in Biotechnology.** - [s.l.]: Elsevier Science Ltd, - 3. - Vol. 13, - pp. 83-85, março de 1995.

HEILIG, Morton L. **Inventor In The Field Of Virtual Reality.** Available at: http://www.mortonheilig.com/InventorVR.html. - s/a. - October 5, 2019. - http://www.mortonheilig.com/InventorVR.html.

JOHNSTON, Angus P. R. et al. Journal to the centre of the cell: Virtual reality immersion into scientific data. **Wiley Online Library.** - [s.l.]: John Wiley & Sons A/S, pp. 105-110, novembro de 2017.

LABSTER, 2019. **Labster -** Bacteria isolation lab.

LAVROFF, Nicholas. **Having fun with Virtual Reality**. Trad. AIELLO Silvia. Rio de Janeiro: Berkeley Brasil Editora, 1994.

LESTON, J. Virtual Reality: the it perspective. **Computer Bulletin,** - Vol. 1, - pp. 12-13, jun de 1996.

LIBARTI, P. L. O.; BARBOSA, V. **Agile Methods**: Monograph. - Limeira/SP: [n.º], 2010.

LIMA, Cláudia Regina Uchôa. The State of the Art of Virtual Relity. **Revista Informática na Educação: Teoria & Prática.** Porto Alegre: [n.º], - 1 : Vol. 2, May 1999.

MAY, Michael; VRIES, Lisbeth Elvira de. Virtual Laboratory Simulation in the Education of Laboratory Technicians-motivation and study intensity. **Biochemistry and Molecular Biology Education** / ed. Biology International Unio of Biochemistry and Molecular. - 3: Vol. 47, pp. 257-262, mai/jun de 2019.

MAKRANSKY, Guido et al. Equivalence of using a desktop virtual reality science simulation at home and in class. **Plos One**, pp. 11-14, abril de 2019.

MANIFESTO, AGILE. Agile Manifesto. **Manifesto for Agile Software Development.** - 2001. - December 15, 2017. Available at: https://agilemanifesto.org/.

MARCONI, M. A.; LAKATOS, E. M. **Methodology of scientific work**. São Paulo : Atlas, 2001. - 6.

MARTINS, Hyram et al. Developing Virtual Reality applications with HTC VIVE in Unity C#. **Wine Science and Technique**, 2017.

MARTINS, Valéria Farinazzo. **Virtual Reality Environments and Applications Development Process**: Dissertation / Computer Science; Universidade Federal de São Carlos. São Carlos : [n.º], 2000.

MENDES, Maximiliano Augusto de Araújo. **Production and use of animations and videos in the teaching of cell biology for the first series of high school**: Dissertation / Institutes of Biological Sciences, Physics and Chemistry; Univerrsidade de Brasília. Brasília : [n.º], 2010, p. 103.

MONQUEIRO, Julio Cesar Bess. **Blender 3D 2.5**: tutorial for beginners, parts 1 to 8 a start to Blender. Available at: https://www.hardware.com.br/. - May 17, 2011. - October 10, 2017. - https://www.hardware.com.br/tutoriais/novo-blender3d-iniciantes/.

MONTEIRO, Edivaldo Antonio. **Use of Agile Techniques in Exclusive Software Testing Projects**: Monograph / Post-Graduation Course in Informatics; Federal University of Paraná. Curitiba: [n.º], 2013, p. 58.

MORAN, José Manuel. **The education we want: new challenges and how to get there.** Campinas: Papirus, 2012. - 5.

MORAN, José Manuel; BACICH, Lilian Cássia. **Active methodologies for an innovative education**: a theoretical and practical approach. Porto Alegre: Penso, 2018, p. 238.

MORAN, José Manuel. Innovative Teaching and Learning with technology support. In: MORAN, José Manuel; MASETTO, Marcos T.; BEHRENS, Marilda Aparecida. **New technologies and pedagogical mediation**. Campinas: Papirus, 2013. - 21.

NELSON, David L.; COX, Michael M. **Lehninger's Principles of Biochemistry**. Porto Alegre: Artmed, 2019. - 7.

NETO, Arilo Claudio Dias. **Software Engineering Magazine** - Introduction to Software Testing. DEVMEDIA. - 2007. - 20 MARCH 2019. Available at: https://www.devmedia.com.br/artigo-engenharia-de-software-introducao-a-teste-de-software/803.

NETTO, Antonio Valerio; MACHADO, Liliane dos Santos; OLIVEIRA, Maria Cristina F. de. **Virtual Reality: foundations and applications**. Florianópolis: VisualBooks, 2002.

OLIVEIRA, E. M.; STOLAAR, H. L. F.; MORAES, K. C. M. Making science teaching (cell biology) more dynamic and effective through practical activities. **Encontro Latino Americano de Inciação Científica**, 13th Latin **American Meeting of** Graduate Studies of the University of Vale do Paraíba. São José dos Campos: [n.º], pp. 1-6, 2009.

ORLANDO, Tereza Cristina et al. Planning, assembly and application of didactic models for approaching cellular and molecular biology in high school by graduate students of biological sciences. **Brazilian Journal of Teaching Biochemistry and Molecular Biology.** - [s.l.] : Alfenas, Vol. 1, 2009.

PALMERO, Maria Luz Rodríguez; MOREIRA, Marco Antonio. Menateles models of the structure and functioning of the cell: case studies. **Investigations in Science Teaching.** Porto Alegre: [n.º], - 2: Vol. 4, 1999.

PALMERO, Maria Luz Rodriguez. Bibliographical review of biology teaching and research in the cell study. **Investigações em Ensino de Ciências,** - 3: Vol. 5, pp. 237-263, 2000.

PRINS, Jan F. et al. A virtual environment for steered molecular dynamics. **Future Generation Computer Systems.** - [s.l.] : Elsevier Science B.V. All rights reserved, - 4. - Vol. 15, pp. 485-495, julho de 1999.

REIMEIER, Tanja; GROPENGIEBER, Harald. On the Roots of Difficulties is Learning about Cell Division: Process-based analysis of studentes conceptual development in teaching experiments. **International Journal of Science Education,** - 7. - Vol. 30, pp. 923-939, junho de 2008.

RHEINGOLD, Howard. **Virtual Reality**: the computer-generated artificial worlds that will change our lives. - [s.l.]: Gedisa, 1994. p. 408.

RIZZATO, Andréia C.; NUNES, Fátima L. S. Virtual Reality applied to education: reflections on the state of the art and the future. **IX Workshop on Virtual and Augmented Reality.** - s/a.

RODRIGUES, Leude Pereira; MOURA, Lucilene Silva; TESTA, Edimácio. The Traditional and the Modern as to the didactics of higher education. **Scientific Journal of ITPAC.** Araguaína: 3 : Vol. 4, pp. 1-9, 2011.

SANTOS, Tessa Larissa de; OLIVEIRA, Diene Eire Melo Bortotti de. The use of teaching resources in higher education. **II Jornada de didática and I CEMAD research seminar.** Londrina: 2013.

SEVERINO, Antônio Joaquim. **Methodology of Scientific Work**. São Paulo: Cortez, 2007. - 23.

SIQUEIRA, Jairo. **Applied Creativity**. Rio de Janeiro: [n.º], 2015. - 1.

SOARES, Michel dos Santos. Comparison between Agile and Traditional Methodologies for Software Development. **INFOCOMP Journal of Computer Science,** - 2: Vol. 3, pp. 8-13, November 2004.

SOTRES, Rogelio Rodríguez et al. Simulated Site-directed Mutations in a Virtual Reality Environment as a Powerful Aid for Teaching the Three-dimensional Structure of Proteins. **Educación Química,** - 4. - Vol. 20, pp. 461-465, outubro de 2009.

TORI, Romero. The presence of technology. In: ______. **Education without Distance**: Interactive technologies in distance reduction in teaching and learning. São Paulo: Senac São Paulo, 2010.

TORI, Romero; KIRNER, Claudio. Fundamentals of Virtual Reality. In: TORI, Romero; KIRNER, Cláudio; SISCOUTTO, Robson. **Fundamentals and technology of virtual and augmented reality**. Porto Alegre: Brazilian Computer Society, 2006.

TORI, Romero. **Education without distance**: interactive technologies in distance reduction in teaching and learning. Sao Paulo: Educational Craftsmanship, 2017. - 2.

Blender Tutorials. Available at: https://www.graphicsandprogramming.net/por. - November 3rd, 2017. - https://www.graphicsandprogramming.net/por/tutorial/blender.

UNITY. Unity 3D. - Unity Technologies ApS, January 16, 2019. - April 8, 2019. - Available at: https://unity.com/pt.

VALÉRIO, NETTO Antonio. Virtual Reality: saves time and money in the automotive chain. **Revista Engenharia Automotiva e Aeroespacial,** - 9, pp. 32-37, April 2002.

VIANNA, Ilca Oliveira de Almeida. **Methodology of scientific work: a didactic focus of scientific production**. - [s.l.]: Editora E.P.U., 2001.

VINCE J. **Virtual Reality Systems**. - [s.l.] : Addison-Wesley, 2004.

VIVE VIVE ™ | Virtual Reality System. Available at: www.vive.com. - April 04, 2018. - https://www.vive.com/us/product/vive-virtual-reality-system/.

WINN, Willian. **A Conceptual Basis for Educational Applications of Virtual Reality**. - agosto de 1993.

XU, Xiaoqian et al. Exploration of an interactive "Virtual and Actual Combined" teaching mode in medical developmental biology. **Biochemistry and Molecular Biology** Education, - 6: Vol. 46, pp. 585-591, novembro de 2018.

ZUANON, Átima Clemente Alves; DINIZ, Raphael Hermano Santos; NASCIMENTO, Luziane Helena do. Construction of didactic games for the teaching of Biology: a resource for the integration of students into teaching practice. **Revista Brasileira de Ensino de Ciência e Tecnologia,** - 3: Vol. 3, pp. 49-59, 2010.

APPENDICES

INFORMED CONSENT FORM*

IDENTIFICATION DATA

Project Title: Virtual Reality Environment Simulator for Plasma Membrane Teaching

Responsible Researchers: Juliardnas Rigamont dos Reis, Prof. Dr. Dionne Cavalcante Monteiro.

Institution of Responsible Researchers: Universidade Federal do Pará - Núcleo de Inovação e Tecnologias Aplicadas a Ensino e Extensão - NITAE2.

Contact: (91) 992434516 - Juliardnas Rigamont dos Reis (juliarigamont@yahoo.com.br).

Student's name:_____________________ ______________________________.

Age: R   .G   ._____________________________________ years   .

Legal guardian (where applicable)_______: _________________________________

R.G. Legal____ guardian: ________

The student ____is being invited to participate in the research project **"Plasma Membrane Simulator in Virtual Reality environment as a didactic resource for the teaching of Cell Biology"**, under the supervision of Prof. Dr. Dionne Cavalcante Monteiro.

This study aims to develop and validate the simulator in a virtual reality environment of the plasma membrane, to be used as a didactic tool in the teaching-learning process of the Cellular Biology discipline, besides investigating how the use of the technological resource influences the understanding of the biological processes worked by the students.

The activity developed consists in the application of the Simulator in Virtual Reality Environment of the Plasma Membrane, developed by the mentioned researchers, in the undergraduate courses that have in their menu the disciplines of Cellular Biology, where the theme explored in the *software* is taught. During the application session of the simulator, the students should navigate through the cell, explore the plasma membrane and the transport of substances that occur through it.

The method of application and analysis of student behaviour will be divided in two moments: application of the simulator in virtual reality of the plasma membrane and application of a questionnaire. The activity will be carried out in class, and all information obtained by the registration or experiment will be confidential.

The performance of the work will not cause any damage either academically or of any nature to the participating student. The experiment aims to validate the tool and investigate the effect it has on the process of teaching and learning the plasmatic membrane.

The participation of the graduate is **voluntary**, and this consent may be withdrawn at any time, without prejudice to any of the parties involved.

It is reiterated that any information generated will be confidential, being of exclusively scientific use. The privacy of the participants of this research will always be preserved and ensured in such a way that the participants are not identified nominally.

Please be advised that you will not pay or be compensated for your participation.

Any eventual doubts about this research can be solved with a phone call or e-mail to the responsible researchers (information available at the beginning of the term), or by the Graduate Program in Creativity and Innovation in Higher Education Methodologies (PPGCIMES) of the Center for Innovation in Technologies Applied to Teaching and Extension (NITAEE) of the Federal University of Pará - Campus Universitário do Guamá, Av. Augusto Corrêa, 01, Guamá, CEP: 66.075-110, Belém-PA, telephone 3201-8698, e-mail: ppgcimes.ufpa@gmail.com.

We appreciate your appreciation of this document and your eventual agreement to participate in this survey.

I, ________________________________, R.G. n° ___________________,
legally responsible for ________________________________, R.G. n°
___________________, declare having been informed and agree to your participation, as a volunteer, in the research project described above.

Bethlehem, ____ of ___________________.

____
Signature of the participant or person in charge

________________________ ____   ________________________________
Witness   **Witness**

*Term of Free Informed Consent presented, according to the rules of resolution 4466/2012 of December 12, 2012.

APPENDIX II - QUESTIONNAIRE FOR UNDERGRADUATE AND POSTGRADUATE STUDENTS

FEDERAL UNIVERSITY OF PARÁ - UFPA **INNOVATION AND EXTENSION TECHNOLOGIES CORE - NITAE2** **POSTGRADUATE PROGRAM CREATIVITY AND INNOVATION IN HIGHER EDUCATION METHODOLOGIES - PPGCIMES** **RESEARCH LINE: METHODOLOGICAL INNOVATIONS IN HIGHER EDUCATION - INOVAMES**	PPGCIMES

Project: Virtual Reality environment simulator for teaching Plasma Membrane.

Gender Identity: Mal   e ☐ Female   Age:________ ☐   Educational   Institution: ________________

Course: Discipline   : ____

THE FOLLOWING ARE SOME ISSUES RELATED TO THE USE OF THE SIMULATOR IN VIRTUAL REALITY OF THE PLASMA MEMBRANE AND ITS INFLUENCE ON THE TEACHING-LEARNING PROCESS. ANSWER THEM USING THE SCALE REPRESENTED IN THE LEGEND BELOW.

.	☐ 1 Strongly disagre e ☐ 2 Strongly disagre e ☐ 3 Indifferent ☐ 4 Agre e ☐ 5 Strongly Agree	
	As for using the simulator, please answer the questions below:	
1.	The simulator is difficult to manipulate	1 2 ☐ 3 ☐ 4 ☐ 5 ☐☐
2.	The instructions provided were sufficient for use and handling of the resource	1 2 ☐ 3 ☐ 4 ☐ 5 ☐☐
3.	It was very interesting to use this resource	1 2 ☐ 3 ☐ 4 ☐ 5 ☐☐
4.	I enjoyed participating in a Virtual Reality experience	1 2 ☐ 3 ☐ 4 ☐ 5 ☐☐
5.	At times I was bored while using the resource	1 2 ☐ 3 ☐ 4 ☐ 5 ☐☐
6.	The use of virtual reality in teaching stimulates learning	1 2 ☐ 3 ☐ 4 ☐ 5 ☐☐
7.	I performed well while using the resource	1 2 ☐ 3 ☐ 4 ☐ 5 ☐☐
8.	After the application of the resource I understood the content better	1 2 ☐ 3 ☐ 4 ☐ 5 ☐☐
9.	Would you like to use this feature other times?	☐YES ☐NO
10.	Would you recommend the use of the simulator to other people?	☐YES ☐NO

11. Did you experience difficulties using the simulator? Comment on your experience, highlighting what your difficulties were

12. What is the issue related to the plasma membrane that became easier to assimilate after using the simulator?

13. The basic structure of all cell membranes is a phospholipidic layer. During the use of the simulator, what were the movements performed by these structures that you were able to visualize?

14. What types of membrane protein did you notice during the use of the simulator? Exemplify and cite differences between them.

15. What types of transport did you notice while using the simulator? Exemplify

16. Ion channels are highly selective, allowing only ions with appropriate size and loads to pass through. Did the simulator allow you to observe this? Exemplify

17. In your opinion, what could be improved in the virtual environment, so that your experience and learning are more positive?

Thank you for your willingness and cooperation in using and evaluating the Virtual Reality Simulator as a teaching resource for teaching Plasma Membrane. Your opinion is very important to help improve this didactic resource.

APPENDIX III - QUESTIONNAIRE FOR CELL BIOLOGY TEACHERS

FEDERAL UNIVERSITY OF PARÁ - UFPA
INNOVATION AND EXTENSION TECHNOLOGIES CORE - NITAE2
POSTGRADUATE PROGRAM CREATIVITY AND INNOVATION IN HIGHER EDUCATION METHODOLOGIES - PPGCIMES
RESEARCH LINE: METHODOLOGICAL INNOVATIONS IN HIGHER EDUCATION - INOVAMES

Project: Virtual Reality environment simulator for teaching Plasma Membrane.

Sex: Mal e ☐ Femal e ☐   Age: _______   **Teaching institution that teaches:** __

After testing the simulator, evaluate it using the table below, where the Likert scale is used, which allows you to classify the statements, present in the first column, on a scale that varies from 1 to 5, which allows you to quantify them, allowing an objective analysis of how the simulator works. You can complement your assessment by answering the open questions.

Teacher analysis on the usability of the Virtual Reality Simulator for teaching Plasma Membrane	I totally disagree	I disagree	Indifferent	I agree	I totally agree
Weights	1	2	3	4	5
Using this simulator can make the teaching-learning process more interesting.					
It is easier to teach Cell Biology using this simulator or others similar to it.					
The simulator can be used to review the subject matter.					
The information presented is useful, objective and contributes to understanding the subject matter.					
The simulator is easy to handle.					
The representations of the interface information are easy to recognize (menu, modeling etc).					
It would be interesting to have simulators like this, addressing other content					

1. Have you ever had contact with any teaching oriented virtual reality resources?

( ) Yes   ( ) No

2. If the answer to the previous question is yes, please describe where and how you experienced it.

3. Based on the virtual reality experience you experienced when using the **Virtual Reality Simulator for the teaching of Plasma Membrane**, do you believe that this technology is a good resource to be used in the teaching of Cellular and Molecular Biology? Why do you think so?

4. Would you adopt this simulator in your cellular and molecular biology classes? Why would you?

5. Are the concepts of molecular and cellular biology addressed in the simulator correct?

6. Regarding the simulator design, do you consider that the formats and colors used in the modeling are appropriate?

( ) Yes ( ) No

7. If they're not, what could improve?

8. Do you think any aspect of the content covered in the simulator needs to be adequate and/or improved? What?

Thank you for your willingness and cooperation in using and evaluating the Virtual Reality Simulator as a teaching resource for teaching Plasma Membrane. Your opinion is very important to help improve this didactic resource

PHOSPHOLIPIDS PUZZLE ASSEMBLY GUIDE

AUTHORS

Juliardnas Rigamont dos Reis - UFPA
CV: http://lattes.cnpq.br/9792261025601451

Ana Cássia Sarmento Ferreira - IFPA
CV: http://lattes.cnpq.br/2022102405472089

Dionne Cavalcante Monteiro - UFPA
CV: http://lattes.cnpq.br/4423219093583221

Hey, player! Are you ready to have fun and learn all about phospholipids? This puzzle will challenge you to build the spatial formulas of phospholipids through didactic models that facilitate the teaching-learning process. But don't worry, this guide is designed to help you. It gathers clues for you to figure out how to assemble each formula. So you can develop meaningful learning by playing!

You will understand the organization of atoms and the chemical bonds between them differently. Instead of simply writing the structural formula or viewing a two-dimensional image of phospholipids, you will have the opportunity to handle and build the formulas yourself, following some tips provided here.

To play, you have at your disposal 425 (four hundred and twenty-five) hydrogen atoms, represented by white balls; 230 (two hundred and thirty) carbon atoms, symbolized by black balls; 50 (fifty) oxygen atoms, reproduced in red balls; 10 (ten) phosphorus atoms, equivalent to green balls, and 10 (ten) nitrogen atoms, demonstrated in blue balls. Such atoms constitute the phospholipids of the plasma membrane of animal cells. In addition, the "phospholipids puzzle" features metal rods to establish the chemical bonds between these atoms.

Now it's up to you! Play, have fun and learn!

WHY DO WE LEARN MORE WHEN WE PLAY?

This guide was developed because the visual stimulus will contribute to your learning, after all, most of the information your brain receives is through your vision. This occurs both in reading texts and in observing images, but image assimilation is much faster because the brain is able to perceive an image in one tenth of a second [1].

The learning process is closely linked to the stimuli sent to the human brain, and depends on them to activate reasoning and assimilation. Therefore, the teacher's work does not consist in motivating the student, but in stimulating his reasoning [2].

But when the teaching model is abstract and vertical, the student's probability to learn less is great, because by assuming a passive posture, he restricts himself to learning by decorating and copying, but does not exercise the ability to create and transform, actions fundamental to his formation [3].

[1] OLIVE TREE, 2011.
[2] ZANATA, 2014.
[3] PACHANE, 2001.

Therefore, it is important that the playful activities are present in the classroom, as structuring elements of the teaching process and triggers significant learning, those in which the human being needs to integrate his abilities to think, act and feel, without hypertrophy. But playfulness in higher education? Yes! Who said that in university education one cannot work ludically? First of all, it is necessary to demystify the restricted understanding of playfulness as recreational or leisure action. Playfulness does include recreation, but it is not limited to it. Recreation is a playful, external activity. It is not playfulness. Leisure is also not synonymous with playfulness [4].

Leisure is the space/time in which playful activities take place. Playfulness is an inner state, an attitude of those who fully experience playfulness. The absence of visual language has reduced teaching-learning, because the practices used are repeatedly academic, or very based on abstract content. Not that the contents are not important. They are our raw material, that is an indisputable question. But it is necessary to understand that the human being does not learn only from the intellect [4].

[4] D'ÁVILA, 2014.

And it was from these principles that the "phospholipid puzzle" was designed for you. When assembling the pieces, you will be able to see what you are learning, becoming more active in the teaching-learning process and facilitating the memorization of the contents. This is because by putting the pieces together you will be able to visualize and understand the chemical bonds established by the atoms that make up the phospholipids.

So, player, the visual stimulus represents a great potential for the teaching-learning process, because, besides arousing your interest and curiosity, it can contribute to better retain the content.

After the phospholipids theory class, you will be able to assemble your own puzzle, a visual tool that will benefit the learning of the biochemistry of this specific molecule.

This puzzle is important for the learning process because it materializes an abstract knowledge. When the student works only with memorization, it is common for them to forget the concepts after class. But when the student does, learning is effective because, according to the learning pyramid, the human being learns: 10% when he reads; 20% when he listens; 30% when he observes; 50% when he sees and hears; 70% when he argues with others; 80% when he does; 95% when he teaches others [5].

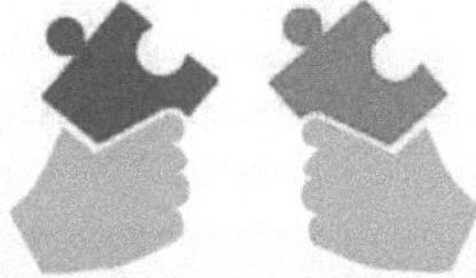

[5] GLASSER 2017.

Phospholipids are the most abundant lipids in the plasma membrane. They are amphiphilic molecules, that is, they have a hydrophilic region, soluble in aqueous media, and a hydrophobic region, insoluble in water, but soluble in organic lipids and solvents. (COOPER, 2007).

All phospholipids consist of two fatty acid chains, which make up the hydrophobic 'tail' of phospholipids (Figure 1), attached to a glycerol or sphingosine molecule, which in turn has another grouping attached to it. This grouping may be ethanolamine, serine, choline or inositol. Glycerol, together with this grouping, makes up the hydrophilic "head" (figure 1).

Figure 1: Model of the phospholipid molecule.

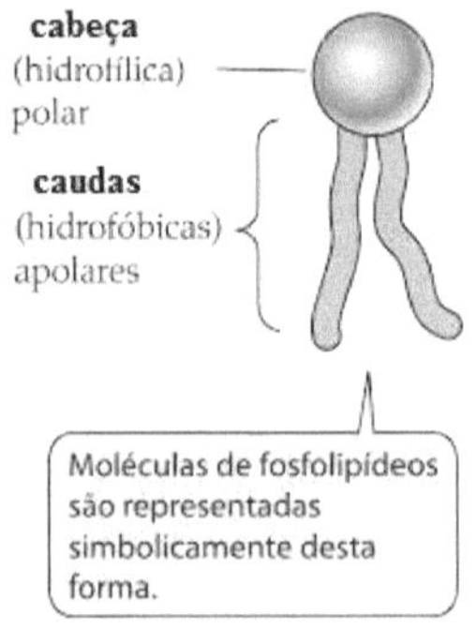

Source: Hiil et al. (2012).

Fatty acids differ from each other by their extensive hydrocarbon chain (figure 2), as well as by the presence, number and position of these double bonds. When they do not have double bonds, they are called "saturated". When there are double bonds, they are called "unsaturated".

Figure 2: Saturated and unsaturated fatty acid.

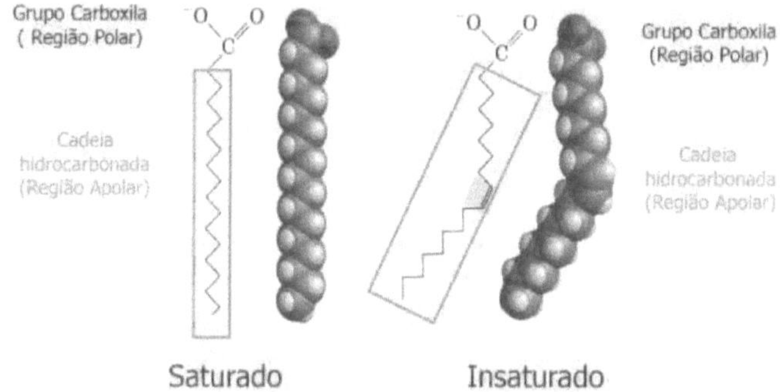

Source: GOULART, Flavia Cristina, s/d.

WHAT TYPES OF PHOSPHOLIPIDS WILL BE ASSEMBLED?

This puzzle represents phosphoglycerides, which have a main three-carbon (C3H5O3) glycerol chain. The carbons adjacent to the glycerol bind to two long hydrocarbon chains, while the third carbon atom of the glycerol binds to a phosphate group. This in turn binds to various types of groups, constituting different phosphoglycerides. The most abundant in the animal plasma membrane are: phosphatidylethanolamine, phosphatidylserine, phosphatidylcholine and phosphatidylinositol (ALBERTS et al., 2017).

Another important class of phospholipids, also represented in this puzzle, are sphingolipids. Instead of glycerol, they present sphingosine, which is a long acetyl chain with an amino group (NH2) and two hydroxyl groups (OH) at one end. In these phospholipids, a hydrocarbon tail is bound to the amino group, and a phosphocholine group is bound to the end hydroxyl group, as is the case with sphingomyelin (ALBERTS et al., 2017).

TIPS FOR PHOSPHOLIPIDS ASSEMBLY

The "tails" of all phospholipids (phosphatidylinositol, phosphatidylcholine, phosphatidylserine, phosphatidylethanolamine and sphingomyelin) represented in this puzzle are symbolized by two hydrocarbon chains and one carboxyl group (COOH or CO_2H) terminal.

Saturated tail: formed by 18 carbons.

Unsaturated Tail: degree of unsaturation 18:1 (Δ9).

Note that the difference between the various types of phospholipids is in the head.

1. Phosphatidylethanolamine

Its head consists of glycerol (C3H8O3), phosphate group ($PO4^{-3}$) and ethanolamine (C_2H7NO). The glycerol binds to the tails and its other end binds to the phosphate group, which also binds to ethanolamine.

2. Phosphatidylserine

Its head is formed by glycerol (C3H8O3), phosphate group ($PO4^{-3}$) and serine (C_2H7NO3). The glycerol joins the tails and its other end joins the phosphate group, which in turn connects with the serine.

3. Phosphatidylcholine

The components of the head are glycerol (C3H8O3), phosphate group (PO4 $^{-3)}$ and choline (C5H14NO). The glycerol binds to the tails and its other end binds to the phosphate group, which binds to the hill.

4. Phosphatidylinositol

The chemical composition of its head is glycerol (C3H8O3), phosphate group (PO4 $^{-3)}$ and inositol (C6H12O6). The glycerol is chemically bound with the tails and its other end is chemically bound with the phosphate group, which in turn binds to inositol.

5. Sphingomyelin

The chemical composition of its head is sphingosine (C6H10O2N), phosphate group (PO4-3) and choline (C5H14NO). The sphingosine is chemically bound to the tails and its other end is bound to the phosphate group, which in turn is bound to the choline.

The "tails" of all phospholipids (phosphatidylinositol, phosphatidylcholine, phosphatidylserine, phosphatidylethanolamine and sphingomyelin) presented in this puzzle are represented by two hydrocarbon chains and one carboxyl group ($COOH$ or CO_2H) terminal.

Saturated tail: $CH3\ (CH2)_{16}CO_2H$

Unsaturated Tail: $CH3\ (CH2)_{7CH} = CH\ (CH2)_{7CO_2H}$

What differentiates one phospholipid from another is the head, and four types of phospholipids have glycerol in their formation, and the chemical representation of glycerol is:

Figure 3: Structural formula of glycerol.

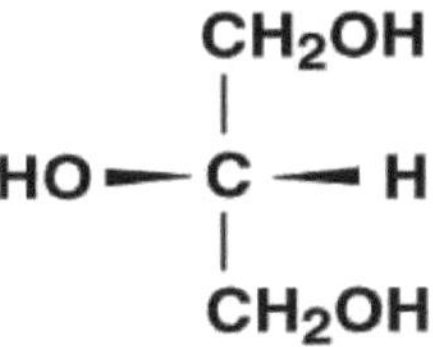

Source: authors, 2019.

1. Phosphatidylethanolamine

The head consists of glycerol (C3H8O3), phosphate group (PO4-3) and ethanolamine (C_2H7NO). The glycerol binds to the tails and its other end binds to the phosphate group, which also binds to ethanolamine.

Figure 4: Structural formula of phosphatidylethanolamine.

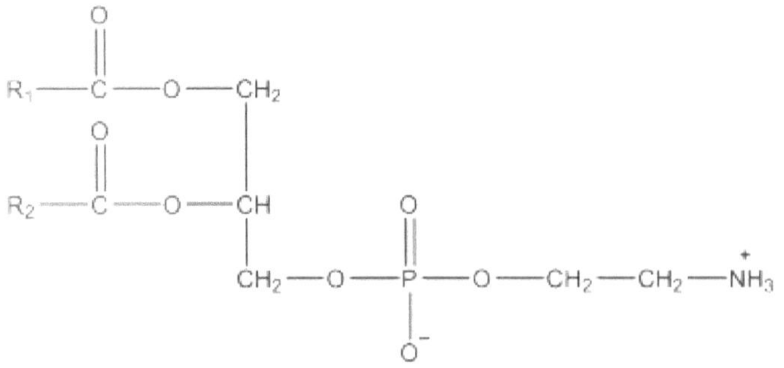

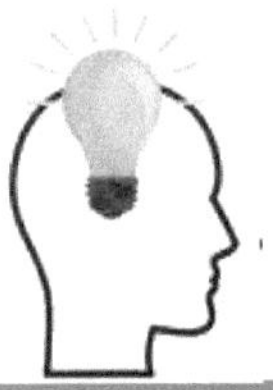

Source: authors, 2019.

2.   Phosphatidylserine

Its head consists of glycerol (C3H8O3), phosphate group (PO4-3) and serine $_{(C_2H7NO3)}$. The glycerol binds to the tails and its other end binds to the phosphate group, which binds to the serine.

Figure 5: Structural formula of phosphatidylserine.

Source: authors, 2019.

3.   Phosphatidylcholine

The components of the head are glycerol (C3H8O3), phosphate group (PO4-3) and choline (C5H14NO). The glycerol binds to the tails and its other end binds to the phosphate group, which binds to the hill.

Figure 6: Structural formula of phosphatidylcholine.

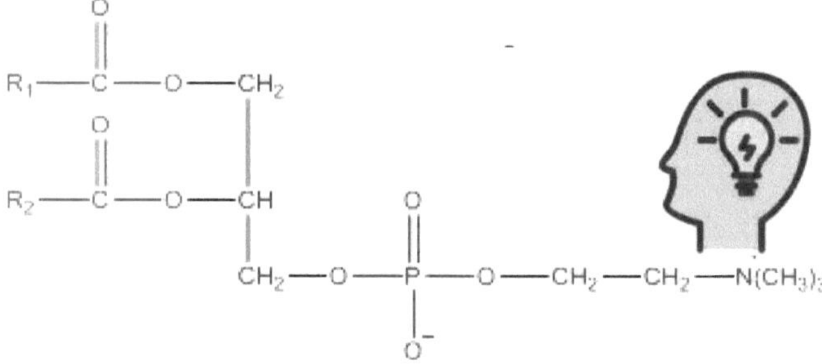

Source: authors, 2019.

4. Phosphatidylinositol

The chemical composition of its head is glycerol (C3H8O3), phosphate group (PO4-3) and inositol (C6H12O6). The glycerol is chemically bound with the tails and its other end is chemically bound with the phosphate group, which in turn binds to inositol.

Figure 7: Structural formula of Phosphatidylinositol

Source: authors, 2019.

5. Sphingomyelin

The chemical composition of its head is sphingosine (C6H10O2N), phosphate group (PO4-3) and choline (C5H14NO). The sphingosine is chemically bound to the tails and its other end is bound to the phosphate group, which in turn is bound to the choline.

Figure 8: Sphingomyelin's structural formula.

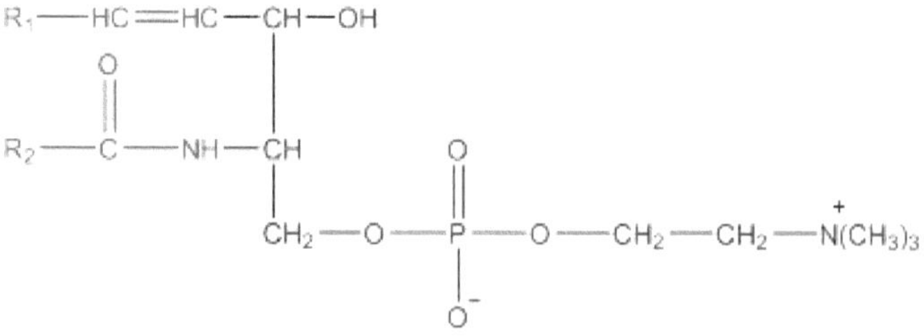

$$R_1 - HC = HC - CH - OH$$

$$R_2 - \overset{\overset{O}{\|}}{C} - NH - CH$$

$$CH_2 - O - \overset{\overset{O}{\|}}{\underset{\underset{O^-}{|}}{P}} - O - CH_2 - CH_2 - \overset{+}{N}(CH_3)_3$$

Source: authors, 2019.

REFERENCES BIBLIOGRAPHICS

ALBERTS, B. et al. **Molecular Cell Biology**. Translation by Ardala Elisa Breda Andrade et al. 6th Ed. Porto Alegre: Artmed, 2017.

COOPER, G.M. **The cell**: A molecular approach. 3rd Ed. Porto Alegre: Artmed, 2007.

D'ÁVILA, Cristina M. **Lúdica Didática**: Saberes Pedagógicos e Ludicidade no Contexto da Educação Superior. September/2014. Available at: https://www. researchgate. net/publication/324848479_Didatica_ludica_saberes_pedagogicos_e_ludicidade_no_contexto_da_educacao _superior> Access on: May 21, 2019.

GLASSER, W. (2017). William Glasser. Source: PPD: Available at: < http://www. ppd. net. br/william-glasser/>. Access on: May 21, 2019.

GOULART, Flavia Cristina, s/d. **Lipids**. Available at: https://www.marilia.unesp.br/Home/Instituicao/Docentes/FlaviaGoulart/lipidios.pdf> Access on: May 21, 2019.

HILL, Richard W.; GORDON, A.; WYSE, Margaret Anderson. **Animal Physiology** (electronic resource). 2nd edition. Electronic data. Porto Alegre: Artmed, 2012.

OLIVEIRA, Gilberto Gonçalves. **Neuroscience and Educational Processes**: A Necessary Knowledge in Teacher Training. Master's program in education research line: culture and educational processes. Uberaba 2011. Available at https://www. researchgate. net/publication/277132829 >. Access on 16 May 2019.

PACHANE, G. The hybrid character of the "Portuguese language" in higher education. In: Anais XIII COLE: **Reading Congress in Brazil**. UNICAMP, 2001.

ZANATA, Milena Hoppen. The Contribution of Stimulation to Learning. **IDEAU Education Magazine**. Vol. 9 - Nº 20 - July - December 2014 Semestral ISSN: 1809-6220. Available at ideau. com. br/getulio/restrito/upload/revistasartigos/222_1.pdf >. Access on 15 May 2019.

More
Books!

OMNIScriptum

Printed by Books on Demand GmbH, Norderstedt / Germany